Joint Publication 1, *Doctrine for the Armed Forces of the United States*, is the capstone publication for all joint doctrine, presenting fundamental principles and overarching guidance for the employment of the Armed Forces of the United States. This represents the evolution in our warfighting guidance and military theory that forms the core of joint warfighting doctrine and establishes the framework for our forces' ability to fight as a joint team.

It is vital that we not only develop our military capabilities, but also strengthen the capacity of other government departments and agencies. This publication ties joint doctrine to the national security strategy and national military strategy and describes the military's role in the development of national policy and strategy. It thus provides the linkage between joint doctrine and the contribution of other government departments and agencies and multinational endeavors.

As we look globally at our posture and the associated strategic risk, it is imperative that our doctrine also rapidly adjust to reflect our wartime footing. The guidance in this publication will enable current and future leaders of the Armed Forces of the United States to organize, train, and execute worldwide missions as our forces transform to meet emerging challenges. The joint force must simultaneously think ahead at the strategic level, stay current at the operational level, and be informed by tactical level developments.

I challenge all commanders to ensure the widest distribution of this capstone joint publication and actively promote the use of all joint publications at every opportunity. I further challenge you to study and understand the guidance contained in this publication and teach these principles to your subordinates. Only then will we be able to fully exploit the remarkable military potential inherent in our joint teams.

MARTIN E. DEMPSEY
General, U.S. Army

PREFACE

1. Scope

This publication is the capstone joint doctrine publication and provides doctrine for unified action by the Armed Forces of the United States. It specifies the authorized command relationships and authority that military commanders can use, provides guidance for the exercise of that military authority, provides fundamental principles and guidance for command and control, prescribes guidance for organizing and developing joint forces, and describes policy for selected joint activities. It also provides the doctrinal basis for interagency coordination and for US military involvement in multiagency and multinational operations.

2. Purpose

a. The US Armed Forces fulfill unique and crucial roles, defending the US against all adversaries while serving the Nation as a bulwark and the guarantor of its security and independence. The US Armed Forces function within the American system of civil-military relations and serve under the civilian control of the President, the Commander in Chief. The US Armed Forces embody the highest values and standards of American society and the profession of arms.

b. The nature of the challenges to the US and its interests demand that the Armed Forces operate as a closely integrated joint team with interagency and multinational partners across the range of military operations. Using a whole-of-government approach is essential to advancing our interests to strengthen security relationships and capacity by, with, and through military forces of partner nations, US and foreign government agencies, state and local government agencies, and intergovernmental or nongovernmental organizations. To succeed, we must refine and proportionally integrate the military with all of the tools of American power and work with our partner nations to do the same. Our military must maintain its conventional superiority while continuing to enhance its capacity to defeat threats. As long as nuclear weapons exist, our nuclear deterrent capability must also be maintained and modernized. When international forces are needed to respond to threats and keep the peace, we will make every effort to ensure international partners are ready, able, and willing. We will continue to build support in other countries and promote global peace and stability through the United Nations and other regional organizations, such as the North Atlantic Treaty Organization and the African Union.

c. Joint Operations. Effective integration of joint forces is intended to address functional or geographic vulnerabilities. This does not mean that all forces will be equally represented in each operation. Joint force commanders (JFCs) may choose the capabilities they need from the forces at their disposal.

Preface

3. Application

a. This publication is written to assist members of the Armed Forces of the United States, including the National Guard, to operate successfully together. The joint team is composed of the members of each Service, Department of Defense agencies, as well as associated civilians supporting governmental and private sector workforces. The guidance in this publication is broad, authoritative, and serves as a foundation for the development of more specific joint guidance. This doctrine will be followed except when, in the judgment of the commander, exceptional circumstances dictate otherwise.

b. To ensure the Armed Forces achieve their fullest potential, all US military leaders shall incorporate the doctrine and philosophy of this publication into their efforts to develop leaders and train forces for joint and multinational operations. JFCs shall incorporate the guidelines and philosophies of this doctrine as fundamental precepts while conducting interagency coordination.

c. The Services and United States Special Operations Command (in areas unique to special operations) have specific responsibilities under Title 10, United States Code (USC), to organize, train, equip, prepare, and maintain their forces. The National Guard has similar, specific responsibilities under Title 32, USC, and includes domestic operations. These forces are employed under JFCs. Service equipment, systems, and manpower skills form the very core of US military capability. Joint warfare relies upon effective coordination of Service capabilities and expertise. When integrated into joint operations with partner military Services and other defense, logistical, and intelligence agencies, they become capable of unified action. Successful joint operations merge capabilities and skill sets of assigned Service components. Interoperability and effective integration of service capabilities enhance joint operations to accomplish US Government objective(s), building on US traditions of conducting joint operations that began with the Revolutionary War.

d. The growing threats to US and allied interests throughout the world demand US forces be proficient across the range of military operations. The fundamental principles that guide operations are recorded in joint doctrine. Joint operations are conducted routinely and efficiently in the current operational environment. To maintain and enhance this efficiency, joint leaders must diligently study, apply, teach, and ultimately provide insights to improve joint doctrine.

SUMMARY OF CHANGES
REVISION OF JOINT PUBLICATION 1, DATED 02 MAY 2007,
CHANGE 1, DATED 20 MARCH 2009

- Adds a theory section to the introductory chapter.

- Adds a joint force development chapter, including a section on joint concepts and assessment.

- Establishes a taxonomy relating to war, warfare, campaign, and operation.

- Establishes a taxonomy relating to policy, strategy, doctrine, and concepts.

- Establishes and defines "global synchronizer."

- Clarifies the role of the Department of Defense relative to information operations to improve efficiency in planning and execution of military operations.

- Expands the role of commander's communication synchronization and information operations.

- Adds information on Global Force Management Implementation Guidance resulting from the closure of Joint Forces Command.

- Introduces "total force fitness" as a value of joint service.

- Reduces redundancies and improves continuity between Joint Publication (JP) 1, *Doctrine for the Armed Forces of the United States,* and JP 3-0, *Joint Operations.*

- Reduces redundancies and improves continuity between JP 1, *Doctrine for the Armed Forces of the United States,* and JP 5-0, *Joint Operation Planning.*

- Establishes information as the seventh joint function. (Change 1)

Intentionally Blank

TABLE OF CONTENTS

EXECUTIVE SUMMARY ix

CHAPTER I
THEORY AND FOUNDATIONS

Section A. Theory I-1
- Fundamentals I-1
- War.. I-2
- Warfare I-4
- Forms of Warfare.. I-5
- Levels of Warfare I-7
- Campaigns and Operations I-9
- Task, Function, and Mission I-9

Section B. Foundations I-10
- Strategic Security Environment and National Security Challenges I-10
- Instruments of National Power and the Range of Military Operations I-12
- Joint Operations I-16
- Joint Functions I-17
- Joint Operation Planning.. I-19
- Law of War I-21

CHAPTER II
DOCTRINE GOVERNING UNIFIED DIRECTION OF ARMED FORCES

- National Strategic Direction II-1
- Strategic Guidance and Responsibilities... II-3
- Unified Action II-8
- Roles and Functions... II-9
- Chain of Command II-9
- Unified Command Plan.. II-11
- Combatant Commands.. II-11
- Military Departments, Services, Forces, Combat Support Agencies, and
  National Guard Bureau II-11
- Relationship Among Combatant Commanders,
  Military Department Secretaries, Service Chiefs, and Forces II-13
- Interagency Coordination... II-13
- Multinational Operations II-21

CHAPTER III
FUNCTIONS OF THE DEPARTMENT OF DEFENSE AND ITS MAJOR
COMPONENTS

Section A. Department of Defense III-1
- General.. III-1

v

Table of Contents

- Organizations in the Department of Defense III-1
- Functions of the Department of Defense III-1
- Functions and Responsibilities Within the Department of Defense III-2
- Executive Agents III-2

Section B. Joint Chiefs of Staff III-3
- Composition and Functions III-3
- Chairman of the Joint Chiefs of Staff III-4
- Vice Chairman of the Joint Chiefs of Staff III-5
- Joint Staff III-6

Section C. Military Departments and Services III-6
- Common Functions of the Services and the
  United States Special Operations Command III-6

Section D. Combatant Commanders III-7
- General III-7
- Geographic Combatant Command Responsibilities III-8
- Functional Combatant Command Responsibilities III-9
- Statutory Command Authority III-11
- Authority Over Subordinate Commanders III-11
- Department of Defense Agencies III-12

CHAPTER IV
   JOINT COMMAND ORGANIZATIONS

Section A. Establishing Unified and Subordinate Joint Commands IV-1
- General IV-1
- Unified Combatant Command IV-5
- Specified Combatant Command IV-9
- Subordinate Unified Command IV-10
- Joint Task Force IV-10

Section B. Commander, Staff, and Components of a Joint Force IV-12
- Commander Responsibilities IV-12
- Staff of a Joint Force IV-13
- Service Component Commands IV-15
- Functional Component Commands IV-17

Section C. Discipline IV-18
- Responsibility IV-18
- Uniform Code of Military Justice IV-19
- Rules and Regulations IV-19
- Jurisdiction IV-19
- Trial and Punishment IV-20

Section D. Personnel Service Support and Administration IV-21
- Morale, Welfare, and Recreation IV-21

Table of Contents

- Awards and Decorations IV-21
- Efficiency, Fitness, and Performance Reports IV-21
- Total Force Fitness IV-22
- Personnel Accountability IV-22
- Religious Affairs IV-22
- Information Management IV-23

CHAPTER V
JOINT COMMAND AND CONTROL

Section A. Command Relationships V-1
- General Principles V-1
- Combatant Command (Command Authority) V-2
- Operational Control V-6
- Tactical Control V-7
- Support V-8
- Support Relationships Between Combatant Commanders V-9
- Support Relationships Between Component Commanders V-10
- Command Relationships and Assignment and Transfer of Forces V-11
- Other Authorities V-12
- Command of National Guard and Reserve Forces V-13

Section B. Command and Control of Joint Forces V-14
- Background V-14
- Command and Control Fundamentals V-14
- Organization for Joint Command and Control V-18
- Joint Command and Staff Process V-19
- Command and Control Support V-19
- National Military Command System V-20
- Nuclear Command and Control System V-20
- Defense Continuity Program V-21

CHAPTER VI
JOINT FORCE DEVELOPMENT

Section A. Fundamentals of Joint Force Development VI-1
- Principles VI-1
- Authorities VI-1

Section B. Joint Force Development Process VI-2
- Joint Force Development VI-2
- Joint Doctrine VI-3
- Joint Education VI-4
- Joint Training VI-6
- Lessons Learned VI-8
- Joint Concepts and Assessment VI-9

vii

Table of Contents

APPENDIX

A   Establishing Directive (Support Relationship) Considerations A-1
B   The Profession of Arms B-1
C   References C-1
D   Administrative Instructions D-1

GLOSSARY

Part I   Abbreviations and Acronyms GL-1
Part II   Terms and Definitions GL-5

FIGURE

I-1   Principles of War I-3
I-2   Levels of Warfare I-7
I-3   Range of Military Operations I-14
II-1   Strategy, Planning, and Resourcing Process II-5
II-2   Unified Action II-8
II-3   Chain of Command II-10
II-4   Notional Joint Interagency Coordination Group Structure II-19
II-5   Notional Composition of a Civil-Military Operations Center II-20
II-6   Notional Multinational Command Structure II-24
III-1   Command Functions of a Combatant Commander III-11
IV-1   Possible Components in a Joint Force IV-3
IV-2   Unified Combatant Command Organizational Options IV-6
IV-3   Specified Combatant Command Organizational Options IV-9
IV-4   Subordinate Unified Command Organizational Options IV-10
IV-5   Joint Task Force Organizational Options IV-11
V-1   Command Relationships Synopsis V-2
V-2   Categories of Support V-10
V-3   Transfer of Forces and Command Relationships Overview V-11
VI-1   Joint Force Development Life Cycle VI-2

EXECUTIVE SUMMARY
COMMANDER'S OVERVIEW

- **Discusses the Theory and Foundations of Joint Doctrine**

- **Characterizes Doctrine Governing Unified Direction of Armed Forces**

- **Outlines the Functions of the Department of Defense and Its Major Components**

- **Details Doctrine for Joint Commands**

- **Describes the Fundamental Principles for Joint Command and Control**

- **Addresses Joint Force Development**

Theory and Foundations

This publication provides overarching guidance and fundamental principles for the employment of the Armed Forces of the United States.

Joint Publication 1 is the capstone publication of the US joint doctrine hierarchy. It is a bridge between policy and doctrine and describes the authorized command relationships and authority that military commanders can use and other operational matters derived from Title 10, United States Code (USC). The purpose of joint doctrine is to enhance the operational effectiveness of joint forces by providing fundamental principles that guide the employment of US military forces toward a common objective.

Jointness of the Joint Force

Jointness implies cross-Service combination wherein the capability of the joint force is understood to be synergistic, with the sum greater than its parts (the capability of individual components). The joint force is a values based organization. The character, professionalism, and values of our military leaders have proven to be vital for operational success.

War is socially sanctioned violence to achieve a political purpose.

War can result from the failure of states to resolve their disputes by diplomatic means. War historically involves nine principles, collectively and classically known as the principles of war (objective, offensive, mass, economy of force, maneuver, unity of command, security, surprise, and simplicity).

Warfare is the mechanism, method, or modality of armed

Warfare continues to change and be transformed by society, diplomacy, politics, and technology. The US

---

ix

Executive Summary

*conflict against an enemy.
It is "the how" of waging war.*

military recognizes two basic forms of warfare—traditional and irregular. The forms of warfare are applied not in terms of an "either/or" choice, but in various combinations to suit a combatant's strategy and capabilities.

The US military recognizes two basic forms of warfare—traditional and irregular.

A useful dichotomy for thinking about warfare is the distinction between traditional and irregular warfare (IW). **Traditional warfare** is characterized as a violent struggle for domination between nation-states or coalitions and alliances of nation-states. With the increasingly rare case of formally declared war, traditional warfare typically involves force-on-force military operations in which adversaries employ a variety of conventional forces and special operations forces (SOF) against each other in all physical domains as well as the information environment (which includes cyberspace). **IW** is characterized as a violent struggle among state and non-state actors for legitimacy and influence over the relevant population(s). In IW, a less powerful adversary seeks to disrupt or negate the military capabilities and advantages of a more powerful military force, which usually serves that nation's established government.

Levels of Warfare

While the various forms and methods of warfare are ultimately expressed in concrete military action, the three levels of warfare—strategic, operational, and tactical—link tactical actions to achievement of national objectives. There are no finite limits or boundaries between these levels, but they help commanders design and synchronize operations, allocate resources, and assign tasks to the appropriate command.

Campaigns and Operations

An **operation** is a sequence of tactical actions with a common purpose or unifying theme. An operation may entail the process of carrying on combat, including movement, supply, attack, defense, and maneuvers needed to achieve the objective of any battle or campaign. A **campaign** is a series of related major operations aimed at achieving strategic and operational objectives within a given time and space.

Task, Function, and Mission

A **task** is a clearly defined action or activity assigned to an individual or organization. It is a specific assignment that must be done as it is imposed by an appropriate authority. A **function** is the broad, general, and

Executive Summary

enduring role for which an organization is designed, equipped, and trained. **Mission** entails the task, together with the purpose, that clearly indicates the action to be taken and the reason therefore.

Strategic Security Environment and National Security Challenges

The **strategic security environment** is characterized by uncertainty, complexity, rapid change, and persistent conflict. This environment is fluid, with continually changing alliances, partnerships, and new national and transnational threats constantly appearing and disappearing. The strategic security environment presents broad national security challenges likely to require the employment of joint forces in the future. The US military will undertake the following activities to deal with these challenges: secure the homeland, win the Nation's wars, deter our adversaries, security cooperation, support to civil authorities, and adapt to changing environment.

Instruments of National Power and the Range of Military Operations

The ability of the US to advance its national interests is dependent on the effectiveness of the United States Government (USG) in employing the instruments of national power to achieve national strategic objectives. The military instrument of national power can be used in a wide variety of ways that vary in purpose, scale, risk, and combat intensity. These various ways can be understood to occur across a continuum of conflict ranging from peace to war. Mindful that the operational level of warfare connects the tactical to the strategic, and operations and campaigns are themselves scalable, the US uses the construct of the range of military operations to provide insight into the various broad usages of military power from a strategic perspective.

Joint Operations

Although individual Services may plan and conduct operations to accomplish tasks and missions in support of Department of Defense (DOD) objectives, the primary way DOD employs two or more Services (from two Military Departments) in a single operation, particularly in combat, is through joint operations. Joint operations is the general term to describe military actions conducted by joint forces and those Service forces in specified command relationships with each other.

Joint Functions

There are significant challenges to effectively integrating and synchronizing Service and combat

xi

Executive Summary

support agency (CSA) capabilities in joint operations. Functionally related capabilities and activities can be grouped. These groupings, which we call *joint functions,* facilitate planning and employment of the joint force. In addition to *command and control (C2),* the joint functions include *intelligence, fires, movement and maneuver, protection, sustainment, and information.*

Joint Operation Planning

Joint operation planning provides a common basis for discussion, understanding, and change for the joint force, its subordinate and higher headquarters, the joint planning and execution community, and the national leadership. In accordance with the **Guidance for Employment of the Force (GEF),** adaptive planning supports the transition of DOD planning from a contingency-centric approach to a strategy-centric approach. The **Adaptive Planning and Execution (APEX)** system facilitates iterative dialogue and collaborative planning between the multiple echelons of command. The combatant commanders' (CCDRs') participation in the Joint Strategic Planning System and APEX system helps to ensure that warfighting and peacetime operational concerns are emphasized in all planning documents.

Joint operation planning is the way the military links and transforms national strategic objectives into tactical actions.

Law of War

It is DOD policy that the Armed Forces of the United States will adhere to the law of war, often called the law of armed conflict, during all military operations. The law of war is the body of law that regulates both the legal and customary justifications for utilizing force and the conduct of armed hostilities; it is binding on the US and its individual citizens.

Doctrine Governing Unified Direction of Armed Forces

National Strategic Direction

National strategic direction is governed by the Constitution, US law, USG policy regarding internationally recognized law, and the national interest as represented by national security policy. This direction leads to unified action. National policy and planning documents generally provide national strategic direction.

xii   JP 1

Executive Summary

Strategic Guidance and Responsibilities

The **national security strategy (NSS)** provides a broad strategic context for employing military capabilities in concert with other instruments of national power.

The **national defense strategy (NDS)**, signed by Secretary of Defense (SecDef), outlines DOD's approach to implementing the President's NSS.

The **National Military Strategy**, signed by the Chairman of the Joint Chiefs of Staff (CJCS), supports the aims of the NSS and implements the NDS. It describes the Armed Forces' plan to achieve military objectives in the near term and provides a vision for maintaining a force capable of meeting future challenges.

The **GEF** provides Presidential and SecDef politico-military guidance. The GEF is guided by the Unified Command Plan (UCP) and NDS and forms the basis for strategic policy guidance, campaign plans, and the Joint Strategic Capabilities Plan.

The **National Response Framework,** developed by the Department of Homeland Security, establishes a comprehensive, national-level, all-hazards, all-discipline approach to domestic incident management.

Unified Action

Unified action synchronizes, coordinates, and/or integrates joint, single-Service, and multinational operations with the operations of other USG departments and agencies, nongovernmental organizations (NGOs), intergovernmental organizations (IGOs) (e.g., the United Nations), and the private sector to achieve **unity of effort.** Unity of command within the military instrument of national power **supports the national strategic direction** through close coordination with the other instruments of national power. The CJCS and all CCDRs are in pivotal positions to facilitate the planning and conduct of unified actions in accordance with the guidance and direction received from the President and SecDef in coordination with other authorities (i.e., multinational leadership).

Roles and Functions

Roles are the broad and enduring purposes for which the Services and the combatant commands (CCMDs) were established in law. **Functions** are the appropriate

xiii

Executive Summary

assigned duties, responsibilities, missions, or tasks of an individual, office, or organization.

Chain of Command

The President and SecDef exercise authority, direction, and control of the Armed Forces through two distinct branches of the chain of C2. One branch runs from the President, through SecDef, to the CCDRs for missions and forces assigned to their commands. For purposes other than the operational direction of the CCMDs, the chain of command runs from the President to SecDef to the Secretaries of the Military Departments and, as prescribed by the Secretaries, to the commanders of Military Service forces. The Military Departments, organized separately, operate under the authority, direction, and control of the Secretary of that Military Department. The Secretaries of the Military Departments exercise administrative control (ADCON) over Service retained forces through their respective Service Chiefs. CCDRs prescribe the chain of command within their CCMDs and designate the appropriate command authority to be exercised by subordinate commanders.

Unified Command Plan

The President, through the UCP, establishes CCMDs. Commanders of unified CCMDs may establish subordinate unified commands when so authorized by SecDef.

Combatant Commands

CCDRs exercise combatant command (command authority) (COCOM) of assigned forces. The CCDR may delegate operational control (OPCON), tactical control (TACON), or establish support command relationships of assigned forces. Unless otherwise directed by the President or SecDef, COCOM may not be delegated.

Military Departments, Services, Forces, Combat Support Agencies, and National Guard Bureau

The Secretaries of the Military Departments are responsible for the administration and support of Service forces. They fulfill their responsibilities by exercising ADCON through the Service Chiefs. Service Chiefs have ADCON for all forces of their Service. Commanders of Service forces are responsible to Secretaries of the Military Departments through their respective Service Chiefs for the administration, training, and readiness of their unit(s). The National Guard Bureau is responsible for ensuring that units and members of the Army National Guard and the Air

xiv

JP 1

National Guard are trained by the states to provide trained and equipped units to fulfill assigned missions in federal and non-federal statuses. In addition to the Services above, a number of DOD agencies provide combat support or combat service support to joint forces and are designated as CSAs. The CSA directors are accountable to SecDef.

Relationship Between Combatant Commanders, Military Department Secretaries, Service Chiefs, and Forces

The Services and United States Special Operations Command (USSOCOM) (in areas unique to special operations [SO]) share the division of responsibility for developing military capabilities for the CCMDs. Unified action demands maximum interoperability. The forces, units, and systems of all Services must operate together effectively, in part through interoperability. CCDRs will ensure maximum interoperability and identify interoperability issues to the CJCS, who has overall responsibility for the joint interoperability program.

Interagency Coordination

Interagency coordination is the cooperation and communication that occurs between departments and agencies of the USG, including DOD, to accomplish an objective. CCDRs and subordinate joint force commanders (JFCs) must consider the potential requirements for interagency, IGO, and NGO coordination as a part of their activities within and outside of their operational areas. Unity of effort can only be achieved through close, continuous interagency and interdepartmental coordination and cooperation, which are necessary to overcome discord, inadequate structure and procedures, incompatible communications, cultural differences, and bureaucratic and personnel limitations.

Multinational Operations

Operations conducted by forces of two or more nations are termed "multinational operations."

Much of the information and guidance provided for unified action and joint operations are applicable to multinational operations. However, differences in laws, doctrine, organization, weapons, equipment, terminology, culture, politics, religion, and language within alliances and coalitions must be considered. Attaining unity of effort through unity of command for a multinational operation may not be politically feasible, but it should be a goal. A coordinated policy, particularly on such matters as multinational force commanders' authority over national logistics (including infrastructure), rules of engagement,

Executive Summary

fratricide prevention, and intelligence, surveillance, and reconnaissance (ISR) is essential for unity of effort.

Functions of the Department of Defense and Its Major Components

Organization in Department of Defense

All functions in the Department of Defense and its component agencies are performed under the authority, direction, and control of the Secretary of Defense (SecDef).

SecDef is the principal assistant to the President in all matters relating to DOD. DOD is composed of the Office of the Secretary of Defense, the Military Departments, the Joint Chiefs of Staff (JCS), the Joint Staff, the CCMDs, the Inspector General, agencies/bureaus, field activities, and such other offices, agencies, activities, and commands established or designated by law, by the President, or by SecDef. As prescribed by higher authority, DOD will maintain and employ Armed Forces to: support and defend the Constitution of the US against all enemies, foreign and domestic; ensure, by timely and effective military action,
the security of the US, its territories, and areas vital to its
interest; and uphold and advance the national policies and interests of the US.

Joint Chiefs of Staff (JCS)

The Joint Staff supports the JCS and constitutes the immediate SecDef military staff.

The JCS consists of the CJCS; the Vice Chairman of the Joint Chiefs of Staff; the Chief of Staff, US Army; the Chief of Naval Operations; the Chief of Staff, US Air Force; the Commandant of the Marine Corps; and the Chief, National Guard Bureau. The CJCS is the principal military advisor to the President, National Security Council, Homeland Security Council, and SecDef.

Common Functions of the Services and the United States Special Operations Command

Subject to the authority, direction, and control of SecDef and subject to the provisions of Title 10, USC, the Army, Marine Corps, Navy, and Air Force, under their respective Secretaries, are responsible for the functions prescribed in detail in Department of Defense Directive 5100.01, *Functions of the DOD and Its Major Components*. USSOCOM is unique among the CCMDs in that it performs certain Service-like functions (in areas unique to SO) (Title 10, USC, Sections 161 and 167).

Combatant Commanders

Geographic combatant commanders (GCCs) are assigned a geographic area of responsibility (AOR) by the President with the advice of SecDef as specified in the UCP. GCCs are responsible for the missions in their

AOR, unless otherwise directed. **Functional combatant commanders (FCCs)** have transregional responsibilities and are normally supporting CCDRs to the GCC's activities in their AOR. FCCs may conduct operations as directed by the President or SecDef, in coordination with the GCC in whose AOR the operation will be conducted. SecDef or Deputy Secretary of Defense may assign a CCDR **global synchronizer** responsibilities. The global synchronizer's role is to align and harmonize plans and recommend sequencing of actions to achieve the strategic end states and objectives of a global campaign plan.

Geographic Combatant Command Responsibilities

Based on the President's UCP, **the Commanders, US Central Command, US European Command, US Pacific Command, US Southern Command, US Africa Command, and US Northern Command, are each assigned a geographic AOR within which their missions are accomplished with assigned and/or attached forces.** Forces under the direction of the President or SecDef may conduct operations from or within any geographic area as required for accomplishing assigned tasks, as mutually agreed by the CCDRs concerned or as specifically directed by the President or SecDef.

Functional Combatant Command Responsibilities

Commander, US Special Operations Command (CDRUSSOCOM) is an FCC who exercises COCOM of all assigned Active Component and mobilized Reserve Component SOF minus US Army Reserve civil affairs and military information support forces. When directed, CDRUSSOCOM provides US-based SOF to a GCC who exercises COCOM of assigned and OPCON of attached SOF through a commander of a theater SO command or a joint SO task force in a specific operational area or to prosecute SO in support of a theater campaign or other operations.

The **Commander, US Strategic Command,** is an FCC who is responsible to:

- Maintain primary responsibility among CCDRs to support the national objective of strategic deterrence;
- Provide integrated global strike planning;
- Synchronize planning for global missile defense;

xvii

Executive Summary

- Plan, integrate, and coordinate ISR in support of strategic and global operations;
- Provide planning, training, and contingent electronic warfare support;
- Synchronize planning for DOD combating weapons of mass destruction;
- Plan and conduct space operations;
- Synchronize planning for cyberspace operations, and
- Provide in-depth analysis and precision targeting for selected networks and nodes.

The **Commander, US Transportation Command,** is an FCC who is responsible to:

- Provide common-user and commercial air, land, and maritime transportation, terminal management, and aerial refueling to support global deployment, employment, sustainment, and redeployment of US forces;
- Serve as the mobility joint force provider;
- Provide DOD global patient movement, in coordination with GCCs, through the Defense Transportation Network; and,
- Serve as the Distribution Process Owner.

Department of Defense Agencies

DOD agencies are organizational entities of DOD established by SecDef under Title 10, USC, to perform a
supply or service activity common to more than one Military Department.

Joint Command Organizations

Establishing Unified and Subordinate Joint Commands

Authority to Establish. In accordance with the National Security Act of 1947 and Title 10, USC, and as described in the UCP, CCMDs are established by the President, through SecDef, with the advice and assistance of the CJCS. Commanders of unified combatant commands may establish subordinate unified commands when so authorized by SecDef through the CJCS. Joint task forces (JTFs) can be established by SecDef, a CCDR, subordinate unified commander, or an existing JTF commander.

xviii   JP 1

Executive Summary

Unified Combatant Command	A unified combatant command is a **command with broad continuing missions under a single** commander and composed of significant assigned components of two or more Military Departments that is established and so designated by the President through SecDef and with the advice and assistance of the CJCS.
Specified Combatant Command	A specified CCMD is a command that has broad continuing missions and is established by the President, through SecDef, with the advice and assistance of the CJCS.
Subordinate Unified Command	When authorized by SecDef through the CJCS, commanders of unified CCMDs may establish subordinate unified commands (also called subunified commands) to conduct operations on a continuing basis in accordance with the criteria set forth for unified CCMDs.
Joint Task Force	A JTF is a joint force that is constituted and so designated by SecDef, a CCDR, a subordinate unified commander, or an existing JTF commander. A JTF may be established on a geographical area or functional basis when the mission has a specific limited objective and does not require overall centralized control of logistics.
Commander Responsibilities	Although specific responsibilities will vary, a JFC possesses the following general responsibilities:

- Provide a clear commander's intent and timely communication of specified tasks, together with any required coordinating and reporting requirements.
- Transfer forces and other capabilities to designated subordinate commanders for accomplishing assigned tasks.
- Provide all available information to subordinate JFCs and component commanders that affect their assigned missions and objectives.
- Delegate authority to subordinate JFCs and component commanders commensurate with their responsibilities.

Staff of a Joint Force	A JFC is authorized to organize the staff and assign responsibilities to individual Service members assigned to the staff as deemed necessary to accomplish assigned missions. The composition of a joint staff should be

Executive Summary

commensurate with the composition of forces and the character of the contemplated operations to ensure that the staff understands the capabilities, needs, and limitations of each element of the force.

Service Component Commands

A Service component command, assigned to a CCDR, consists of the Service component commander and the Service forces (such as individuals, units, detachments, and organizations, including the support forces) that have been assigned to that CCDR. Forces assigned to CCDRs are identified in the Global Force Management Implementation Guidance (GFMIG) signed by SecDef.

Functional Component Commands

JFCs have the authority to establish functional component commands to control military operations. JFCs may decide to establish a functional component command to integrate planning; reduce their span of control; and/or significantly improve combat efficiency, information flow, unity of effort, weapon systems management, component interaction, or control over the scheme of maneuver.

Discipline

The JFC is responsible for the discipline of military personnel assigned to the joint organization. Each Service component in a CCMD is responsible for the discipline of that Service's component forces, subject to Service regulations and directives established by the CCDR. The Uniform Code of Military Justice is federal law, as enacted by Congress; it provides the basic law for discipline of the Armed Forces of the United States. Matters that involve more than one Service and that are within the jurisdiction of the JFC may be handled either by the JFC or by the appropriate Service component commander. Matters that involve only one Service should be handled by the Service component commander, subject to Service regulations.

Joint Command and Control

Command is central to all military action, and unity of command is central to unity of effort.

Inherent in command is the authority that a military commander lawfully exercises over subordinates including authority to assign missions and accountability for their successful completion. **Although commanders may delegate authority to accomplish missions, they may not absolve themselves of the responsibility for the attainment of these missions.** Authority is never absolute; the extent

xx   JP 1

Executive Summary

of authority is specified by the establishing authority, directives, and law.

Combatant Command (Command Authority)

COCOM provides full authority for a CCDR to perform those functions of command over assigned forces involving organizing and employing commands and forces, assigning tasks, designating objectives, and giving authoritative direction over all aspects of military operations, joint training (or in the case of USSOCOM, training of assigned forces), and logistics necessary to accomplish the missions assigned to the command.

Operational Control

OPCON is the command authority that may be exercised by commanders at any echelon at or below the level of CCMD and may be delegated within the command. OPCON is able to be delegated from and lesser authority than COCOM. It is the authority to perform those functions of command over subordinate forces involving organizing and employing commands and forces, assigning tasks, designating objectives, and giving authoritative direction over all aspects of military operations and joint training necessary to accomplish the mission.

Tactical Control

TACON is an authority over assigned or attached forces or commands, or military capability or forces made available for tasking, that is limited to the detailed direction and control of movements and maneuvers within the operational area necessary to accomplish assigned missions or tasks assigned by the commander exercising OPCON or TACON of the attached force. TACON is able to be delegated from and lesser authority than OPCON and may be delegated to and exercised by commanders at any echelon at or below the level of CCMD.

Support

There are four categories of support that a combatant commander may exercise over assigned or attached forces to ensure the appropriate level of support is provided to accomplish

Support is a command authority. A support relationship is established by a common superior commander between subordinate commanders when one organization should aid, protect, complement, or sustain another force. Support may be exercised by commanders at any echelon at or below the CCMD level. The designation of supporting relationships is important as it conveys priorities to commanders and staffs that are planning or executing joint operations. The support command relationship is, by design, a somewhat vague but very flexible arrangement. The

xxi

Executive Summary

mission objectives. They are: general support, mutual support, direct support, and close support.

establishing authority (the common JFC) is responsible for ensuring that both the supported commander and supporting commanders understand the degree of authority that the supported commander is granted.

Support Relationships Between Combatant Commanders

SecDef establishes support relationships between the CCDRs for the planning and execution of joint operations. This ensures that the supported CCDR receives the necessary support.

Support Relationships Between Component Commanders

The JFC may establish support relationships between component commanders to facilitate operations. Component commanders should establish liaison with other component commanders to facilitate the support relationship and to coordinate the planning and execution of pertinent operations.

Command Relationships and Assignment and Transfer of Forces

All forces under the jurisdiction of the Secretaries of the Military Departments (except those forces necessary to carry out the functions of the Military Departments as noted in Title 10, USC, Section 162) are assigned to CCMDs or Commander, United States Element, North American Aerospace Defense Command, or designated as Service retained by SecDef in the GFMIG. A force assigned or attached to a CCMD, or Service retained by a Service Secretary, may be transferred from that command to another CCDR only when directed by SecDef and under procedures prescribed by SecDef and approved by the President.

Other Authorities

ADCON is the direction or exercise of authority over subordinate or other organizations with respect to administration and support, including organization of Service forces, control of resources and equipment, personnel management, logistics, individual and unit training, readiness, mobilization, demobilization, discipline, and other matters not included in the operational missions of the subordinate or other organizations. **Coordinating authority** is the authority delegated to a commander or individual for coordinating specific functions and activities involving forces of two or more Military Departments, two or more joint force components, or two or more forces of the same Service (e.g., joint security coordinator exercises coordinating authority for joint security area operations among the component commanders). **Direct liaison authorized** is that authority granted by a commander (any level) to a

xxii   JP 1

Executive Summary

subordinate to directly consult or coordinate an action with a command or agency within or outside of the granting command.

Command of National Guard and Reserve Forces

When mobilized under Title 10, USC, authority, command of National Guard and Reserve forces (except those forces specifically exempted) is assigned by SecDef to the CCMDs. Those forces are available for operational missions when mobilized for specific periods in accordance with the law or when ordered to active duty and after being validated for employment by their parent Service. Normally, National Guard forces are under the commands of their respective governors in Title 32, USC, or state active duty status.

Command and Control of Joint Forces

Command is the most important role undertaken by a JFC. C2 is the means by which a JFC synchronizes and/or integrates joint force activities. C2 ties together all the operational functions and tasks and applies to all levels of war and echelons of command.

Command and Control Fundamentals

C2 enhances the commander's ability to make sound and timely decisions and successfully execute them. Unity of effort over complex operations is made possible through decentralized execution of centralized, overarching plans or via mission command. Unity of command is strengthened through adherence to the following C2 tenets: clearly defined authorities, roles, and relationships; mission command; information management and knowledge sharing; communication; timely decision making; coordination mechanisms; battle rhythm discipline; responsive, dependable, and interoperable support systems; situational awareness; and mutual trust.

Organization for Joint Command and Control

Component and supporting commands' organizations and capabilities must be integrated into a joint organization that enables effective and efficient joint C2. The JFC should be guided in this effort by the following principles: simplicity, span of control, unit integrity, and interoperability.

Joint Command and Staff Process

The nature, scope, and tempo of military operations continually changes, requiring the commander to make new decisions and take new actions in response to these changes. This may be viewed as part of a cycle, which is repeated when the situation changes significantly.

Executive Summary

Although the scope and details will vary with the level and function of the command, the purpose is constant: analyze the situation and need for action; determine the course of action (COA) best suited for mission accomplishment; and carry out that COA, with adjustments as necessary, while continuing to assess the unfolding situation.

Command and Control Support

A C2 support system, which includes interoperable supporting communications systems, is the JFC's principal tool used to collect, transport, process, share, and protect data and information. To facilitate the execution and processes of C2, military communications systems must furnish rapid, reliable, and secure information throughout the chain of command.

National Military Command System

The **National Military Command System** provides the means by which the President and SecDef can receive warning and intelligence so that accurate and timely decisions can be made, the resources of the Military Services can be applied, military missions can be assigned, and direction can be communicated to CCDRs or the commanders of other commands.

Nuclear Command and Control System

The **Nuclear Command and Control System** supports the Presidential nuclear C2 of the CCMDs in the areas of integrated tactical warning and attack assessment, decision making, decision dissemination, and force management and report back.

Defense Continuity Program

The **Defense Continuity Program** is an integrated program composed of DOD policies, plans, procedures, assets, and resources that ensures continuity of DOD component mission-essential functions under all circumstances, including crisis, attack, recovery, and reconstitution.

Joint Force Development

Principles of Joint Force Development

Joint force development entails the purposeful preparation of individual members of the Armed Forces (and the units that they comprise) to present a force capable of executing assigned missions. It includes joint doctrine, joint education, joint training, joint lessons learned, and joint concept development and assessment.

Executive Summary

Authorities

Joint force development involves synergistic execution of the legislated authorities of the CJCS, the Service Chiefs, and others (such as CDRUSSOCOM). US law (Title 10, USC, Section 153) gives the CJCS authority regarding joint force development, specifically providing authority to develop doctrine for the joint employment of the Armed Forces, and to formulate policies for the joint training of the Armed Forces to include polices for the military education and training of members of the Armed Forces.

Joint Force Development

Joint force development is a knowledge-based enterprise. A discussion of each of the force development subordinate processes follows.

Joint Doctrine

Joint doctrine consists of the fundamental principles that guide the employment of US military forces in coordinated action toward a common objective. It provides the authoritative guidance from which joint operations are planned and executed.

Joint Education

Education is a key aspect of the joint force development process. **Joint education** is the aspect of professional military education that focuses on imparting joint knowledge and attitudes. Joint education can be broadly parsed into three categories: joint professional military education; enlisted joint professional military education; and other joint education.

Joint Training

Joint training prepares joint forces or joint staffs to respond to strategic, operational, or tactical requirements considered necessary by the CCDRs to execute their assigned or anticipated missions. Joint training encompasses both joint training of individuals as well as collective training of joint staffs, units, and the Service components of joint forces.

Lessons Learned

The **joint lessons learned** component of joint force development entails collecting observations, analyzing them, and taking the necessary steps to turn them into "learned lessons"—changes in behavior that improve the mission ready capabilities of the joint force. Properly assessed, these positive and negative observations help senior leaders identify and fix problems, reinforce success, and inside the joint force

Executive Summary

development perspective, adjust the azimuth and interaction of the various lines of effort.

Joint Concepts and Assessment

Joint concepts examine military problems and propose solutions describing how the joint force, using military art and science, may operate to achieve strategic goals within the context of the anticipated future security environment. Joint concepts lead to military capabilities, both non-materiel and materiel, that significantly improve the ability of the joint force to overcome future challenges. A **joint assessment** is an analytical activity based on unbiased trials conducted under controlled conditions within a representative environment, to validate a concept, hypothesis, discover something new, or establish knowledge. Results of an assessment are reproducible and provide defensible analytic evidence for joint force development decisions

CONCLUSION

This publication is the capstone joint doctrine publication and provides doctrine for unified action by the Armed Forces of the United States. It specifies the authorized command relationships and authority that military commanders can use, provides guidance for the exercise of that military authority, provides fundamental principles and guidance for C2, prescribes guidance for organizing and developing joint forces, and describes policy for selected joint activities. It also provides the doctrinal basis for interagency coordination and for US military involvement in multiagency and multinational operations.

CHAPTER I
THEORY AND FOUNDATIONS

"Doctrine provides a military organization with a common philosophy, a common language, a common purpose, and a unity of effort."

General George H. Decker, US Army Chief of Staff, 1960-1962

SECTION A. THEORY

1. Fundamentals

a. This publication provides overarching guidance and fundamental principles for the employment of the Armed Forces of the United States. It is the capstone publication of the US joint doctrine hierarchy and it provides an overview for the development of other joint service doctrine publications. It is a bridge between policy and doctrine and describes authorized command relationships and authority that military commanders use and other operational matters derived from Title 10, United States Code (USC).

b. The purpose of joint doctrine is to enhance the operational effectiveness of joint forces by providing fundamental principles that guide the employment of US military forces toward a common objective. With the exception of Joint Publication (JP) 1, joint doctrine will not establish policy. However, the use of joint doctrine standardizes terminology, training, relationships, responsibilities, and processes among all US forces to free joint force commanders (JFCs) and their staffs to focus their efforts on solving strategic, operational, and tactical problems. Using historical analysis of the employment of the military instrument of national power in operations and contemporary lessons, these fundamental principles represent what is taught, believed, and advocated as what works best to achieve national objectives.

c. As a nation, the US wages war employing all instruments of national power—diplomatic, informational, military, and economic. The President employs the Armed Forces of the United States to achieve national strategic objectives. The Armed Forces of the United States conduct military operations as a joint force. "Joint" connotes activities in which elements of two or more Military Departments participate. Joint matters relate to the integrated employment of US military forces in joint operations, including matters relating to:

(1) National military strategy (NMS).

(2) Deliberate and crisis action planning.

(3) Command and control (C2) of joint operations.

(4) Unified action with Department of Defense (DOD) and interagency partners. The capacity of the Armed Forces of the United States to operate as a cohesive joint team is a key advantage in any operational environment. Unity of effort facilitates decisive

I-1

Chapter I

unified action focused on national objectives and leads to common solutions to national security challenges.

d. **Jointness and the Joint Force.** The Armed Forces of the United States have embraced "jointness" as their fundamental organizing construct at all echelons. Jointness implies cross-Service combination wherein the capability of the joint force is understood to be synergistic, with the sum greater than its parts (the capability of individual components). Some shared military activities are less joint than are "common;" in this usage "common" simply means mutual, shared, or overlapping capabilities or activities between two or more Services.

(1) Fundamentally, joint forces require high levels of interoperability and systems that are conceptualized and designed with joint architectures and acquisition strategies. This level of interoperability reduces technical, doctrinal, and cultural barriers that limit the ability of JFCs to achieve objectives. The goal is to employ joint forces effectively across the range of military operations (ROMO).

(2) All Service components contribute their distinct capabilities to the joint force; however, their interdependence is critical to overall joint effectiveness. Joint interdependence is the purposeful reliance by one Service on another Service's capabilities to maximize complementary and reinforcing effects of both (i.e., synergy), the degree of interdependence varying with specific circumstances.

(3) The synergy that results from the operations of joint forces maximizes the capability of the force. The JFC has the operational authority and responsibility to tailor forces for the mission at hand, selecting those that most effectively and efficiently ensure success.

(4) The joint force is a values based organization. The character, professionalism and values of our military leaders have proven to be vital for operational success. See Appendix B, "Character, Professionalism, and Values," for an expanded discussion of values.

For a more detailed explanation of the ROMO, see paragraph 9, "Instruments of National Power and the Range of Military Operations."

2. War

a. War can result from failure of states to resolve their disputes by diplomatic means. Some philosophers see it as an extension of human nature. Thomas Hobbes stated that man's nature leads him to fight for personal gain, safety, or reputation. Thucydides said nearly the same thing in a different order, citing fear, honor, and interest as the common causes for interstate conflict.

b. Individuals, groups, organizations, cultures, and nations all have interests. Inevitably, some of those interests conflict with the interests of other individuals, groups, organizations, cultures, and nations. Nearly all international and interpersonal relationships are based on power and self-interests manifested through politics. Nations exercise their power through diplomatic, informational, military, and economic means. All forms of statecraft are important, but as the conflicts approach the requirement for the use of force to achieve that nation's interests, military means become predominant and war can result. The emergence of non-state actors has not changed this concept. Non-state actors may not use statecraft as established; however, they do coerce and threaten the diplomatic power of other nations and have used force, terrorism, or support to insurgency to compel a government to act or refrain from acting in a particular situation or manner or to change the government's policies or organization.

c. War is socially sanctioned violence to achieve a political purpose. War historically involves nine principles, collectively and classically known as the principles of war (see Figure I-1). The basic nature of war is immutable, although warfare evolves constantly.

The application of these classic principles in the conduct of joint operations is amplified and expanded in JP 3-0, Joint Operations.

d. As an integral aspect of human culture, war has been defined and discussed in myriad contexts. As an element of statecraft, it has groundings in US and international law and treaty. Classic scholars such as Carl von Clausewitz and Sun Tzu provide valuable perspectives necessary to inform a more complete understanding of the nature of war.

e. Clausewitz believed that war is a subset of the larger theory of conflict. He defined war as a "duel on a larger scale," "an act of force to compel our enemy," and a "continuation of politics by other means." Distilled to its essence, war is a violent struggle between two (or more) hostile and independent wills, each trying to impose itself on the other. As Clausewitz states, "war is a violent clash of wills."

(1) Clausewitz believed that war is characterized by the shifting interplay of a trinity of forces—emotion (irrational), chance (nonrational), and reason (rational)—connected by principal actors that comprise a social trinity of the people, the military forces, and the government.

(2) Clausewitz noted that the conduct of war combines friction, chance, and uncertainty. These variables often combine to cause "the fog of war." These observations remain true today and place a burden on the commander to remain responsive, versatile, and adaptive in real time to create and seize opportunities and reduce vulnerabilities.

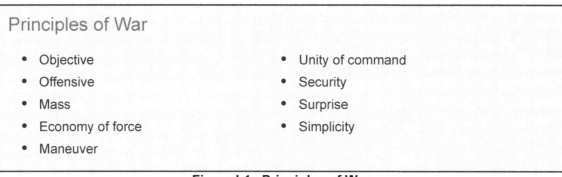

Figure I-1. Principles of War

Chapter I

f. According to Sun Tzu, war is "a matter vital to the state; the province of life or death; the road to survival or ruin." To assess its essentials, he suggests that one analyze it in terms of five fundamental factors: moral influence, weather, terrain, command, and doctrine. He further posits that "what is of supreme importance in war is to attack the enemy's strategy."

g. **Strategy in War.** The two fundamental strategies in the use of military force are strategy of annihilation and strategy of erosion.

(1) The first is to make the enemy helpless to resist us, by physically destroying his military capabilities. This has historically been characterized as annihilation or attrition. It requires the enemy's incapacitation as a viable military force. We may use force when we seek a political objective, such as the overthrow of the enemy leaders. We may also use this strategy in pursuit of more limited political objectives, if we believe the enemy will continue to resist as long as means to do so remain.

(2) The second approach is to convince the enemy that accepting our terms will be less painful than continuing to aggress or resist. This can be characterized as erosion, using military force to erode the enemy leadership's or the enemy society's political will. In such an approach, we use military force to raise the costs of resistance higher than the enemy is willing to pay. We use force in this manner in pursuit of limited political goals that we believe the enemy leadership will ultimately be willing to accept.

(3) Particularly at the higher levels, waging war should involve the use of all instruments of national power that one group can bring to bear against another (diplomatic, informational, military, and economic). While the military focuses on the use of military force, we must not consider it in isolation from the other instruments of national power. Paragraph 9, "Instruments of National Power and the Range of Military Operations," discusses the instruments of national power.

3. Warfare

Warfare is the mechanism, method, or modality of armed conflict against an enemy. It is "the how" of waging war. Warfare continues to change and be transformed by society, diplomacy, politics, and technology.

a. Historian John Keegan offers that war is a universal phenomenon whose form and scope is defined by the society that wages it. The changing form and scope of warfare gives value to delineating the distinction between war and warfare.

b. Understanding the changing nature of warfare frames the context in which wars are fought. Context helps combatants make informed choices as to such essential matters as force structure, force preparation, the conduct of campaigns and operations, and rules of engagement (ROE).

c. The US military recognizes two basic forms of warfare—traditional and irregular. The delineating purpose of each is the strategic focal point of each form. As war is a duality, warfare generally has both traditional and irregular dimensions and offensive and

Theory and Foundations

defensive aspects. The forms of warfare are applied not in terms of an "either/or" choice, but in various combinations to suit a combatant's strategy and capabilities.

4. Forms of Warfare

a. **Overview.** A useful dichotomy for thinking about warfare is the distinction between traditional and irregular warfare (IW). Each serves a fundamentally different strategic purpose that drives different approaches to its conduct; this said, one should not lose sight of the fact that the conduct of actual warfare is seldom divided neatly into these subjective categories. Warfare is a unified whole, incorporating all of its aspects together, traditional and irregular. It is, in fact, the creative, dynamic, and synergistic combination of both that is usually most effective.

> **Note: It is recognized that the symmetry between the naming conventions of traditional and irregular warfare is not ideal. Several symmetrical pair sets—regular/irregular, traditional/nontraditional (or untraditional), and conventional/unconventional—were considered and discarded. Generating friction in the first two instances was the fact that most US operations since the 11 September 2001 terrorist attacks have been irregular; this caused the problem of calling irregular or nontraditional what we do routinely. In the last instance, conventional/unconventional had previous connotation and wide usage that could not be practically overcome.**

b. **Traditional Warfare.** This form of warfare is characterized as a violent struggle for domination between nation-states or coalitions and alliances of nation-states. This form is labeled as traditional because it has been the preeminent form of warfare in the West since the Peace of Westphalia (1648) that reserved for the nation-state alone a monopoly on the legitimate use of force. The strategic purpose of traditional warfare is the imposition of a nation's will on its adversary nation-state(s) and the avoidance of its will being imposed upon us.

(1) In the traditional warfare model, nation-states fight each other for reasons as varied as the full array of their national interests. Military operations in traditional warfare normally focus on an adversary's armed forces to ultimately influence the adversary's government. With the increasingly rare case of formally declared war, traditional warfare typically involves force-on-force military operations in which adversaries employ a variety of conventional forces and special operations forces (SOF) against each other in all physical domains as well as the information environment (which includes cyberspace).

(2) Typical mechanisms for victory in traditional warfare include the defeat of an adversary's armed forces, the destruction of an adversary's war-making capacity, and/or the seizure or retention of territory. Traditional warfare is characterized by a series of offensive, defensive, and stability operations normally conducted against enemy centers of gravity. Traditional warfare focuses on maneuver and firepower to achieve operational and ultimately strategic objectives.

I-5

Chapter I

(3) Traditional warfare generally assumes that the majority of people indigenous to the operational area are not belligerents and will be subject to whatever political outcome is imposed, arbitrated, or negotiated. A fundamental military objective is to minimize civilian interference in military operations.

(4) The traditional warfare model also encompasses non-state actors who adopt conventional military capabilities and methods in service of traditional warfare victory mechanisms.

(5) The near-term results of traditional warfare are often evident, with the conflict ending in victory for one side and defeat for the other or in stalemate.

c. **Irregular Warfare.** This form of warfare is characterized as a violent struggle among state and non-state actors for legitimacy and influence over the relevant population(s). This form is labeled as irregular in order to highlight its non-Westphalian context. The strategic point of IW is to gain or maintain control or influence over, and the support of, a relevant population.

(1) IW emerged as a major and pervasive form of warfare although it is not a historical form of warfare. In IW, a less powerful adversary seeks to disrupt or negate the military capabilities and advantages of a more powerful military force, which usually serves that nation's established government. The less powerful adversaries, who can be state or non-state actors, often favor indirect and asymmetric approaches, though they may employ the full range of military and other capabilities in order to erode their opponent's power, influence, and will. Diplomatic, informational, and economic methods may also be employed. The weaker opponent could avoid engaging the superior military forces entirely by attacking nonmilitary targets in order to influence or control the local populace. Irregular forces, to include partisan and resistance fighters in opposition to occupying conventional military forces, are included in the IW formulation. Resistance and partisan forces, a form of insurgency, conduct IW against conventional occupying powers. They use the same tactics as described above for the weaker opponent against a superior military force to increase their legitimacy and influence over the relevant populations.

(2) Military operations alone rarely resolve IW conflicts. For the US, which will always wage IW from the perspective of a nation-state, whole-of-nation approaches where the military instrument of power sets conditions for victory are essential. Adversaries waging IW have critical vulnerabilities to be exploited within their interconnected political, military, economic, social, information, and infrastructure systems.

(3) An enemy using irregular methods will typically endeavor to wage protracted conflicts in an attempt to exhaust the will of their opponent and its population. Irregular threats typically manifest as one or a combination of several forms including insurgency, terrorism, disinformation, propaganda, and organized criminal activity based on the objectives specified (such as drug trafficking and kidnapping). Some will possess a range of sophisticated weapons, C2 systems, and support networks that are typically characteristic of a traditional military force. Both sophisticated and less sophisticated

irregular threats will usually have the advantages derived from knowledge of the local area and ability to blend in with the local population.

d. To address these forms of warfare, joint doctrine is principally based on a combination of offensive, defensive, and stability operations. The predominant method or combination depends on a variety of factors, such as capabilities and the nature of the enemy.

5. **Levels of Warfare**

a. **General.** While the various forms and methods of warfare are ultimately expressed in concrete military action, the three levels of warfare—strategic, operational, and tactical—link tactical actions to achievement of national objectives (see Figure I-2). There are no finite limits or boundaries between these levels, but they help commanders design and synchronize operations, allocate resources, and assign tasks to the appropriate command. The strategic, operational, or tactical purpose of employment depends on the nature of the objective, mission, or task.

b. **Strategic Level.** Strategy is a prudent idea or set of ideas for employing the instruments of national power in a synchronized and integrated fashion to achieve theater and multinational objectives. At the strategic level, a nation often determines the national (or multinational in the case of an alliance or coalition) guidance that addresses strategic objectives in support of strategic end states and develops and uses national resources to

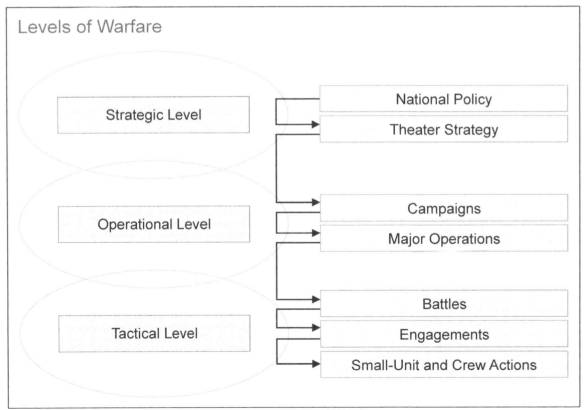

Figure I-2. Levels of Warfare

Chapter I

achieve them. The President, aided by the National Security Council (NSC) and Homeland Security Council (HSC) as the National Security Staff, establishes policy and national strategic objectives. The day-to-day work of the NSC and HSC is accomplished by the combined National Security Staff, the President's principal staff for national security issues. The Secretary of Defense (SecDef) translates these into strategic military objectives that facilitate identification of the military end state and theater strategic planning by the combatant commanders (CCDRs). CCDRs usually participate in strategic discussions with the President and SecDef through the Chairman of the Joint Chiefs of Staff (CJCS) and with partner nations. The CCDR's strategy is an element that relates to both US national strategy and operational-level activities within the theater.

c. **Operational Level.** The operational level links strategy and tactics by establishing operational objectives needed to achieve the military end states and strategic objectives. It sequences tactical actions to achieve objectives. The focus at this level is on the planning and execution of operations using operational art: the cognitive approach by commanders and staffs—supported by their skill, knowledge, experience, creativity, and judgment—to develop strategies, campaigns, and operations to organize and employ military forces by integrating ends, ways, and means. JFCs and component commanders use operational art to determine when, where, and for what purpose major forces will be employed and to influence the adversary's disposition before combat. Operational art governs the deployment of those forces and the arrangement of battles and major operations to achieve operational and strategic objectives.

d. **Tactical Level.** Tactics is the employment and ordered arrangement of forces in relation to each other. The tactical level of war is where battles and engagements are planned and executed to achieve military objectives assigned to tactical units or joint task forces (JTFs). Activities at this level focus on the ordered arrangement and maneuver of combat elements in relation to each other and enemy to achieve combat objectives. An engagement can include a wide variety of activities between opposing forces normally in a short-duration action. A battle consists of a set of related engagements involving larger forces than used in engagements and normally affect the course of an operation or a campaign. Forces at the tactical level generally employ various tactics to achieve their military objectives.

e. While the traditional separate levels of war, as shown in Figure I-2, may help commanders visualize a logical arrangement of missions, allocate resources, and assign tasks to the appropriate command, campaigns and major operations then provide the framework within which the joint force accomplishes the mission; the actual execution is more complicated. With today's constant 24-hour media coverage and easy access to the Internet by our enemies for propaganda, a tactical-level plan and resulting action can have severe operational or strategic implications. For example, an action by one Soldier, Marine, Sailor, or Airman on the battlefield at the tactical level could potentially cause significant disruption to operational and strategic-level planning. Conversely, operations at all levels can be positively influenced by pervasive media coverage, which must be incorporated in plans at all levels. In this sense, during execution all three levels overlap. Commanders and their staffs at all levels must anticipate how their plans, operations, and actions may impact the other levels (those above and those below).

I-8

JP 1

Theory and Foundations

6. Campaigns and Operations

a. Tactics, techniques, and procedures are the fundamental building blocks of concrete military activity. Broadly, actions generate effects; they change in the environment or situation. Tactical actions are the component pieces of operations.

b. An operation is a sequence of tactical actions with a common purpose or unifying theme. An operation may entail the process of carrying on combat, including movement, supply, attack, defense, and maneuvers needed to achieve the objective of any battle or campaign. However, an operation need not involve combat. A major operation is a series of tactical actions, such as battles, engagements, and strikes, conducted by combat forces coordinated in time and place, to achieve strategic or operational objectives in an operational area.

c. A campaign is a series of related major operations aimed at achieving strategic and operational objectives within a given time and space. Planning for a campaign is appropriate when contemplated military operations exceed the scope of a single major operation. Thus, campaigns are often the most extensive joint operations in terms of time and other resources. Some operations can be executed in a single operation and not require campaigning. A noncombatant evacuation, for example, may be executed in a single operation.

7. Task, Function, and Mission

It is worthwhile to discuss three key terms, task, function, and mission, which are relevant to the conduct of warfare at all levels.

a. A task is a clearly defined action or activity assigned to an individual or organization. It is a specific assignment that must be done as it is imposed by an appropriate authority. Function and mission implicitly involve things to be done, or tasks. It is, however, important to delineate between an organization's function and its mission.

b. A function is the broad, general, and enduring role for which an organization is designed, equipped, and trained. Organizationally, functions may be expressed as a task, a series of tasks, or in more general terms. Broadly, a function is the purpose for which an organization is formed. In the context of employing a joint force, joint functions are seven basic groups of related capabilities and activities—C2, intelligence, fires, movement and maneuver, protection, sustainment, and information—that help JFCs integrate, synchronize, and direct joint operations.

c. Mission entails the task, together with the purpose, that clearly indicates the action to be taken and the reason therefore. A mission always consists of five parts: the who (organization to act), what (the task to be accomplished and actions to be taken), when (time to accomplish the task), where (the location to accomplish the task), and why (the purpose the task is to support). Higher headquarters commanders typically assign a mission or tasks to their subordinate commanders, who convert these to a specific mission statement through mission analysis. A mission is what an organization is directed to do.

Chapter I

Functions are the purposes for which an organization is formed. The two are symbiotic. Tasks are relevant to both functions and missions.

For more information on the functions of DOD, refer to Chapter III, "Functions of the Department of Defense and Its Major Components." For more information on joint functions, refer to JP 3-0, Joint Operations.

SECTION B. FOUNDATIONS

"Pure military skill is not enough. A full spectrum of military, para-military, and civil action must be blended to produce success. The enemy uses economic and political warfare, propaganda, and naked military aggression in an endless combination to oppose a free choice of government and suppress the rights of the individual by terror, by subversion, and by force of arms. To win in this struggle, our officers and [Service] men must understand and combine the political, economic, and civil actions with skilled military efforts in the execution of the mission."

President John F. Kennedy
Letter to the United States Army, 11 April 1962

8. Strategic Security Environment and National Security Challenges

a. The strategic security environment is characterized by uncertainty, complexity, rapid change, and persistent conflict. This environment is fluid, with continually changing alliances, partnerships, and new national and transnational threats constantly appearing and disappearing. While it is impossible to predict precisely how challenges will emerge and what form they might take, we can expect that uncertainty, ambiguity, and surprise will dominate the course of regional and global events. In addition to traditional conflicts to include emerging peer competitors, significant and emerging challenges continue to include irregular threats, adversary propaganda, and other information activities directly targeting our civilian leadership and population, catastrophic terrorism employing weapons of mass destruction (WMD), and other threats to disrupt our ability to project power and maintain its qualitative edge.

b. The strategic security environment presents broad national security challenges likely to require the employment of joint forces in the future. They are the natural products of the enduring human condition, but they will exhibit new features in the future. All of these challenges are national problems calling for the application of all the instruments of national power. The US military will undertake the following activities to deal with these challenges:

(1) **Secure the Homeland.** Securing the US homeland is the Nation's first priority. The US homeland is continuously exposed to the possibility of harm from hostile states, groups, and individuals. The Nation must be vigilant and guard against such threats. Defense of the homeland is DOD's highest priority with the goal to identify and defeat threats as far away from the homeland as possible. Deterrence and security cooperation are relevant to homeland defense (HD) and qualify as distinct security challenges.

I-10
JP 1

Theory and Foundations

(2) **Win the Nation's Wars.** Deterring our adversaries is a US goal. Winning the Nation's wars remains the preeminent justification for maintaining capable and credible military forces in the event that deterrence should fail. In the future, as in the past, war and warfare may take a variety of forms. It may erupt among or between states or non-state entities with war-making capabilities. It may manifest as traditional warfare or IW. When the US commits military forces into conflict, success is expected.

(3) **Deter Our Adversaries.** Defending national interests requires being able to prevail in conflict and taking preventive measures to deter potential adversaries who could threaten the vital interests of the US or its partners. These threats could range from direct aggression to belligerent actions that nonetheless threaten vital national interests. Deterrence influences potential adversaries not to take threatening actions. It requires convincing those adversaries that a contemplated action will not achieve the desired result by fear of the consequences. Deterrence is a state of mind brought about by the existence of a credible threat of unacceptable counteraction. Because of the gravity of potential nuclear aggression by a growing list of actors, maintaining a credible nuclear deterrent capability will remain a critical national security imperative.

(4) **Security Cooperation.** Security cooperation encompasses all DOD interactions with foreign defense establishments to build defense relationships that promote specific US security interests, develop allied and friendly military capabilities for self-defense and multinational operations, and provide US forces with peacetime and contingency access to a host nation (HN). Establishing, maintaining, and enhancing security cooperation among our partner nations is important to strengthen the global security framework of the US and its partners. Security cooperation allows us to proactively take advantage of opportunities and not just react to threats. Contributing to security cooperation activities is a large part of what the US military does and will continue to do. Supporting security cooperation activities is an essential element of the CCDR's day-to-day work to enhance regional security and thereby advance national interests. Like deterrence, security cooperation activities can reduce the chances of conflict, but unlike deterrence, it does not involve the threat of force. Security cooperation and deterrence should be complementary as both contribute to security and prevent conflict.

(5) **Support to Civil Authorities.** The US will continue to respond to a variety of civil crises to relieve human suffering and restore civil functioning, most often in support of civil authorities. These crises may be foreign or domestic and may occur independently, as in a natural disaster disrupting an otherwise functioning society, or they may occur within the context of a conflict, such as widespread suffering in a nation embroiled in an insurgency.

(6) **Adapt to Changing Environment.** The strategic security environment and national security challenges are always changing. The ability to address the changing environment and meet our security challenges falls to the instruments of national power and the ability of the Armed Forces of the United States to conduct military operations worldwide.

I-11

Chapter I

9. Instruments of National Power and the Range of Military Operations

a. The ability of the US to advance its national interests is dependent on the effectiveness of the United States Government (USG) in employing the instruments of national power to achieve national strategic objectives. The appropriate governmental officials, often with NSC direction, normally coordinate the employment of instruments of national power.

(1) **Diplomatic.** Diplomacy is the principal instrument for engaging with other states and foreign groups to advance US values, interests, and objectives, and to solicit foreign support for US military operations. Diplomacy is a principal means of organizing coalitions and alliances, which may include states and non-state entities, as partners, allies, surrogates, and/or proxies. The Department of State (DOS) is the USG lead agency for foreign affairs. The credible threat of force reinforces, and in some cases, enables the diplomatic process. Geographic combatant commanders (GCCs) are responsible for aligning military activities with diplomatic activities in their assigned areas of responsibility (AORs). The chief of mission, normally the US ambassador, and the corresponding country team are normally in charge of diplomatic-military activities in a country abroad. In these circumstances, the chief of mission and the country team or another diplomatic mission team may have complementary activities (employing the diplomatic instrument) that do not entail control of military forces, which remain under command authority of the GCC.

(2) **Informational.** Information remains an important instrument of national power and a strategic resource critical to national security. Previously considered in the context of traditional nation-states, the concept of information as an instrument of national power extends to non-state actors—such as terrorists and transnational criminal groups— that are using information to further their causes and undermine those of the USG and our allies. DOD operates in a dynamic age of interconnected global networks and evolving social media platforms. Every DOD action that is planned or executed, word that is written or spoken, and image that is displayed or relayed, communicates the intent of DOD, and by extension the USG, with the resulting potential for strategic effects.

(a) DOD makes every effort to synchronize, align, and coordinate communication activities to facilitate an understanding of how the planning and execution of DOD strategies, plans, operations, and activities will be received or understood by key audiences. This effort is undertaken to improve the efficacy of these actions and create, strengthen, or preserve conditions favorable to advancing defense and military objectives. Communication synchronization entails focused efforts to create, strengthen, or preserve conditions favorable for the advancement of national interests, policies, and objectives by understanding and engaging key audiences through the use of coordinated programs, plans, themes, messages, and products synchronized with the actions of all instruments of national power. In support of these efforts, commanders and staffs at all levels should identify and understand key audience perceptions and possible reactions when planning and executing operations. This understanding of key audience perceptions and reactions is a vital element of every theater campaign and contingency plan. Real or perceived differences between actions and words (the "say-do" gap) are addressed and actively mitigated as appropriate,

Theory and Foundations

since this divergence can directly contribute to reduced credibility and have a negative impact on the ability to successfully execute current and future missions. Attention paid to commander's communication guidance during planning and execution improves the alignment of multiple lines of operation and lines of effort over time and space, which aligns the overarching message with our actions and activities.

(b) Commander's communication guidance is a fundamental component of national strategic direction. It also is essential to our ability to achieve unity of effort through unified action with our interagency partners and the broader interorganizational community. Fundamental to this effort is the premise that key audience beliefs, perceptions, and behavior are crucial to the success of any strategy, plan, and operation. Through commander's communication synchronization (CCS), public affairs (PA), information operations (IO), and defense support to public diplomacy are realized as communication supporting capabilities. Leaders, planners, and operators at all levels need to understand the desired effects and anticipate potential undesired effects of our actions and words, identify key audiences, and when appropriate, actively address their perspectives. Inconsistencies between what US forces say and do can reduce DOD credibility and negatively affect current and future missions. An effective combination of themes, messages, images, and actions, consistent with higher-level guidance, is essential to effective DOD operations.

(c) Within DOD, JFCs implement higher-level communication guidance through the CCS process. JFCs provide guidance and their staffs develop the approach for achieving information-related objectives and ensuring the integrity and consistency of themes, messages, images, and actions to the lowest level through the integration and synchronization of relevant information-related capabilities. Considering the messages our words, images, and actions communicate is integral to military planning and operations and should be coordinated and synchronized with DOD's interorganizational partners.

See JP 3-0, Joint Operations, *and JP 5-0,* Joint Operation Planning, *for more information on commander's communication guidance implementation.*

(3) **Military.** The US employs the military instrument of national power at home and abroad in support of its national security goals. The ultimate purpose of the US Armed Forces is to fight and win the Nation's wars. Fundamentally, the military instrument is coercive in nature, to include the integral aspect of military capability that opposes external coercion. Coercion generates effects through the application of force (to include the threat of force) to compel an adversary or prevent our being compelled. The military has various capabilities that are useful in non-conflict situations (such as in foreign relief). Regardless of when or where employed, the Armed Forces of the United States abide by US values, constitutional principles, and standards for the profession of arms.

(4) **Economic.** A strong US economy with free access to global markets and resources is a fundamental engine of the general welfare, the enabler of a strong national defense. In the international arena, the Department of the Treasury works with other USG agencies, the governments of other nations, and the international financial institutions to

Chapter I

encourage economic growth, raise standards of living, and predict and prevent, to the extent possible, economic and financial crises.

b. The routine interaction of the instruments of national power is fundamental to US activities in the strategic security environment. The military instrument's role increases relative to the other instruments as the need to compel a potential adversary through force increases. The USG's ability to achieve its national strategic objectives depends on employing the instruments of national power discussed herein in effective combinations and all possible situations from peace to war.

c. At the President's direction through the interagency process, military power is integrated with other instruments of national power to advance and defend US values, interests, and objectives. To accomplish this integration, the Armed Forces interact with the other departments and agencies to develop a mutual understanding of the capabilities, limitations, and consequences of military and civilian actions. They also identify the ways in which military and nonmilitary capabilities best complement each other. The NSC plays key roles in the integration of all instruments of national power, facilitating Presidential direction, cooperation, and unity of effort (unified action).

d. Political and military leaders must consider the employment of military force in operations characterized by a complex, interconnected, and global operational environment that affect the employment of capabilities and bear on the decisions of the commander. The addition of military force to coerce an adversary should be carefully integrated with the other instruments of national power to achieve our objectives.

e. The military instrument of national power can be used in a wide variety of ways that vary in purpose, scale, risk, and combat intensity. These various ways can be understood to occur across a continuum of conflict ranging from peace to war. Inside this continuum, it is useful from a strategic perspective to delineate the use of the military

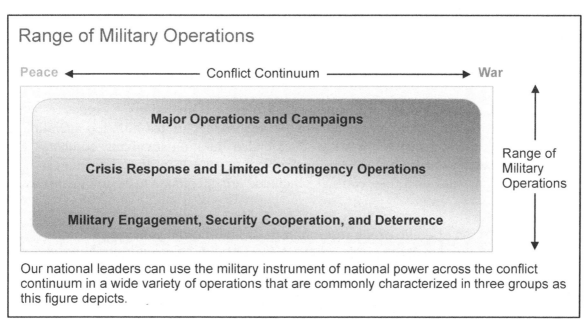

Figure I-3. Range of Military Operations

instrument of national power into three broad categories. Mindful that the operational level of warfare connects the tactical to the strategic, and operations and campaigns are themselves scalable, the US uses the construct of the ROMO to provide insight into the various broad usages of military power from a strategic perspective. See Figure I-3 for these three broad categories, noting that the delineations between the categories are not precise, as each application of military power has unique contextual elements. Each category will be discussed in turn.

(1) **Military Engagement, Security Cooperation, and Deterrence.** These ongoing activities establish, shape, maintain, and refine relations with other nations. Many of these activities occur across the conflict continuum, and will usually continue in areas outside the operational areas associated with ongoing limited contingency operations, major operations, and campaigns.

(a) Military engagement is the routine contact and interaction between individuals or elements of the Armed Forces of the United States and those of another nation's armed forces, domestic or foreign civilian authorities or agencies to build trust and confidence, share information, and coordinate mutual activities.

(b) Security cooperation involves all DOD interactions with foreign defense establishments to build defense relationships that promote specific US security interests, develop allied and friendly military capabilities for self-defense and multinational operations, and provide US forces with peacetime and contingency access to an HN. This includes activities such as security assistance. Security cooperation is a key element of global and theater shaping operations.

(c) Deterrence helps prevent adversary action through the presentation of a credible threat of counteraction. As discussed previously, deterrence convinces adversaries not to take threatening actions by influencing their decision making.

(d) Military actions such as nation assistance (e.g., foreign internal defense, security assistance, humanitarian and civic assistance), counterinsurgency, DOD support to counterdrug operations, show of force operations, and combating WMD activities are applied to meet military engagement, security cooperation, and deterrence objectives.

(2) **Crisis Response and Limited Contingency Operations.** A crisis response or limited contingency operation can be a single small-scale, limited-duration operation or a significant part of a major operation of extended duration involving combat. The associated general strategic and operational objectives are to protect US interests and prevent surprise attack or further conflict. Included are operations to ensure the safety of

---

Note: Some specific crisis response or limited contingency operations may not involve large-scale combat, but could be considered major operations/campaigns depending on their scale and duration (e.g., Operation UNIFIED ASSISTANCE tsunami and Hurricane Katrina relief efforts in 2005, Operation TOMODACHI Japanese tsunami and nuclear relief efforts in 2011).

Chapter I

American citizens and US interests while maintaining and improving US ability to operate with multinational partners to deter the hostile ambitions of potential aggressors (e.g., Operation SHINING EXPRESS in 2003; United States European Command [USEUCOM] launched a joint operation that rescued US citizens and embassy personnel from Monrovia and supported African peacekeeping forces during the Liberian civil war). Many such operations involve a combination of military forces and capabilities in close cooperation with interorganizational partners.

(3) **Major Operations and Campaigns.** When required to achieve national strategic objectives or protect national interests, the US national leadership may decide to conduct a major operation or campaign involving large-scale combat. In such cases, the general goal is to prevail against the enemy as quickly as possible, conclude hostilities, and establish conditions favorable to the US and its interorganizational partners. Major operations and campaigns feature a balance among offensive, defensive, and stability operations through six phases: shape, deter, seize initiative, dominate, stabilize, and enable civil authority. The immediate goal of stability operations often is to provide the local populace with security, restore essential services, and meet humanitarian needs. The long-term goal may be to develop the following: indigenous capacity for securing essential services, a viable market economy, rule of law, democratic institutions, and a robust civil society. Major operations and campaigns typically are composed of multiple phases.

10. Joint Operations

a. In the context of the military instrument of national power, operations are military actions or the carrying out of a strategic, operational, tactical, service, training, or administrative military missions. Operations include combat when necessary to achieve objectives at all levels of war. Although individual Services may plan and conduct operations to accomplish tasks and missions in support of DOD objectives, the primary way DOD employs two or more Services (from two Military Departments) in a single operation, particularly in combat, is through joint operations.

b. Joint operations is the general term to describe military actions conducted by joint forces and those Service forces in specified command relationships with each other. A joint force is one composed of significant elements, assigned or attached, of two or more Military Departments operating under a single JFC.

c. The extensive capabilities available to forces in joint operations enable them to accomplish tasks and missions across the conflict continuum in operations that can range from routine military *engagement* commonly associated with peacetime to large-scale *combat* required to fight and win our Nation's wars. In conjunction with these two extremes, military forces can provide *security* in a wide variety of circumstances and can help other partners restore essential civil services through *relief and reconstruction* in the wake of combat, breakdown of civil order, or natural disaster. These four broad areas, often integrated and adapted to the commander's requirements in a joint operation, represent the military instrument's contribution to meeting our Nation's challenges in the strategic security environment.

I-16   JP 1

Theory and Foundations

(1) The scope and nature of military engagement activities can vary, reflecting differing strategic relationships between the US and partner nations. Engagement includes stability operations and other missions, tasks, and actions that improve the capabilities of, or cooperation with, allies and other partners. It is the primary military contribution to the national challenge of establishing cooperative security. Military engagement may be conducted complementary to broader diplomatic or economic activities, to aid a government's own security activities, and even during war itself. However, commanders and staff must be aware of myriad laws and regulations governing everything from limits on funding and the deployment of military personnel to legislative restrictions on the tasks to which military assistance may be applied. Thus close and continuous coordination between the military and other departments and agencies is essential.

(2) Our Nation may resort to combat when diplomacy or deterrence fails. The fundamental purpose of combat is to defeat armed enemies during traditional warfare, IW, or a combination. It concludes when the mission is accomplished. Combat includes the combination of *offensive* and *defensive* operations and missions to achieve objectives. Combat missions can vary in scale from individual strikes to extensive campaigns and can employ the full range of capabilities available to the military instrument of national power.

(3) Military forces are also adept at providing **security** in a wide range of circumstances. Security missions and tasks cover stability operations, civil support, and other requirements to protect and control civil populations and territory, whether friendly, hostile, or neutral. They also include inherent offensive and defensive measures to protect the joint force. Security actions ultimately seek to reassure rather than compel. Effective security requires a visible and enduring presence. Joint forces can improve security through *security force assistance,* which enhances the capabilities and capacities of a partner nation or regional security organization through training, equipment, advice, and assistance.

(4) A fourth broad area, **relief and reconstruction,** includes stability operations, civil support, and other missions and tasks that restore essential civil services in the wake of combat, a breakdown of civil order, or a natural disaster. The military provides **support to DOS** to assist its relief and reconstruction efforts. Relief and reconstruction assistance may be required in a wide range of situations, such as military occupation, counterinsurgency, and humanitarian crises. Such assistance may be required whether or not civilian relief assets are present and may involve significant civilian contractor support.

11. Joint Functions

a. There are significant complexities to effectively integrating and synchronizing Service and combat support agency (CSA) capabilities in joint operations. These challenges are not new, and they present themselves with consistency. For example, simply getting the joint force to form and deploy in a coherent and desired manner requires integration of organization, planning, and communication capabilities and activities. But to fully employ the joint force in extensive and complex operations requires a much greater array of capabilities and procedures to help the commander and staff integrate and synchronize the joint force's actions. These types of activities and capabilities center on

I-17

Chapter I

the commander's ability to employ the joint force and are grouped under one functional area called *command and control*. In a similar manner, many other functionally related capabilities and activities can be grouped. These groupings, we call *joint functions*, facilitate planning and employment of the joint force.

b. In addition to ***command and control***, the joint functions include ***intelligence, fires, movement and maneuver, protection, sustainment, and information***. Some functions, such as command and control, intelligence, and information apply to all operations. Others, such as fires, apply as the mission requires. A number of subordinate tasks, missions, and related capabilities help define each function, and some apply to more than one joint function. Balancing their complementary but competing processes and capabilities is central to leadership and command of joint operations.

c. The commander must exercise all the joint functions to effectively operate the force and generate combat power. Inadequate integration and balancing of these functions can undermine the cohesion, effectiveness, and adaptability of the force. For example, inattention to protection can deplete combat power unnecessarily, thereby undermining reserves and degrading the force's ability to capitalize on an opportunity or respond to an unforeseen problem. Likewise, inattention to intelligence can leave the force with inadequate information to support decision making or identify opportunities in time to exploit them. Each of the joint functions is discussed below.

(1) *Command and control* encompasses the exercise of authority, responsibility, and direction by a commander over assigned and attached forces to accomplish the mission. *Command* at all levels is the art of motivating and directing people and organizations into action to accomplish missions. *Control* is inherent in command. To *control* is to manage and direct forces and functions consistent with a commander's command authority. Control of forces and functions helps commanders and staffs compute requirements, allocate means, and integrate efforts. Mission command is the preferred method of exercising C2. A complete discussion of tenets, organization, and processes for effective C2 is provided in Section B, "Command and Control of Joint Forces," of Chapter V "Joint Command and Control."

(2) *Intelligence* helps commanders and staffs understand the operational environment and achieve information superiority. Intelligence identifies enemy capabilities and vulnerabilities, projects probable intentions and actions, and is a critical aspect of the joint planning process and execution of operations. It provides assessments that help the commander decide which forces to deploy; when, how, and where to deploy them; and how to employ them in a manner that accomplishes the mission.

(3) *Fires.* To employ fires is to use available weapons and other systems to create a specific lethal or nonlethal effect on a target. Joint fires are those delivered during the employment of forces from two or more components in coordinated action to produce desired results in support of a common objective. Fires typically produce destructive effects, but some ways and means, such as electronic attack and other nonlethal capabilities, can be employed with little or no associated physical destruction.

I-18   JP 1

Theory and Foundations

(4) *Movement and maneuver* encompasses the disposition of joint forces to conduct operations by securing positional advantages before or during execution. This function includes moving or deploying forces into an operational area and maneuvering them within the timeline and to the operational depth necessary to achieve objectives. It uses organic and supporting means and methods that allow a commander to choose where and when to engage an adversary or take best advantage of geographic and environmental conditions.

(5) The *protection* function focuses on conserving the joint force's fighting potential through *active defensive measures* that protect the joint force from an adversary's attack; *passive defensive measures* that make friendly forces, systems, and facilities difficult to locate, strike, and destroy; *technology and procedures* that reduce the risk of fratricide; and *emergency management and response* to reduce the loss of personnel and capabilities due to accidents, health threats, and natural disasters. When directed, the JFC's mission for protection may extend beyond force protection to encompass protection of US civilians; the forces, systems, and civil infrastructure of friendly nations; and our interorganizational partners.

(6) *Sustainment* is the provision of logistics and personnel services necessary to maintain and prolong operations until mission accomplishment. Sustainment provides the JFC flexibility, endurance, and the ability to extend operational reach. Effective sustainment determines the depth to which the joint force can conduct decisive operations, allowing the JFC to seize, retain, and exploit the initiative.

(7) The *information* function encompasses the management and application of information and its deliberate integration with other joint functions to influence relevant-actor perceptions, behavior, action or inaction, and human and automated decision making. The information function helps commanders and staffs understand and leverage the pervasive nature of information, its military uses, and its application during all military operations. This function provides JFCs the ability to integrate the generation and preservation of friendly information while leveraging the inherent informational aspects of all military activities to achieve the commander's objectives and attain the end state.

d. Joint functions should be balanced and integrated with due consideration of competing resources, multiple versus single support capabilities, shifting operational priorities, and differences among Service component practices.

For a more detailed discussion of joint functions, refer to JP 3-0, Joint Operations.

12. Joint Operation Planning

a. Joint operation planning is the way the military links and transforms national strategic objectives into tactical actions. It ties the military instrument of national power to the achievement of national security goals and objectives and is essential in promoting and securing desired global strategic end states during peacetime and war. Planning begins with the end state in mind, providing a unifying purpose around which actions and resources are focused.

Chapter I

b. Joint operation planning provides a common basis for discussion, understanding, and change for the joint force, its subordinate and higher headquarters, the joint planning and execution community (JPEC), and the national leadership. In accordance with the Guidance for Employment of the Force (GEF), adaptive planning supports the transition of DOD planning from a contingency-centric approach to a strategy-centric approach. The Adaptive Planning and Execution (APEX) system facilitates iterative dialogue and collaborative planning between the multiple echelons of command. The APEX system ensures that the military instrument of national power is employed in accordance with national priorities and policy. It also guarantees the plans are rapidly updated and adapted as the situation requires according to changes in policy, strategic guidance, resources, and/or the operational environment. Joint operation planning also identifies capabilities outside of DOD required for accomplishment of strategic end states and objectives and provides a forum for interagency dialogue, coordination, and action.

c. The pursuit and attainment of the US national strategic objectives in today's complex environment requires critical and creative thinking about the challenges facing the joint force. Joint operation planning fosters understanding, allowing commanders and staff to clearly understand the operational environment and identify the problem(s) and problem framework to enable further detailed planning. The planning process, both iterative and collaborative, facilitates the development of options to effectively meet the complex challenges facing joint forces throughout the world.

d. The body of information and understanding created during planning allows CCDRs and their subordinate JFCs and their staffs to adapt to uncertain and changing environments and to anticipate and rapidly act in crisis situations. Joint operation planning produces multiple options to employ the US military and to integrate US military actions with other instruments of US national power in time, space, and purpose to achieve global strategic end states. Planning also identifies and aligns resources with military actions, providing a framework to identify and mitigate risk. The CCDRs' participation in the Joint Strategic Planning System (JSPS) and APEX system helps to ensure that warfighting and peacetime operational concerns are emphasized in all planning documents.

e. Joint operation planning is fundamental to assessing risk and identifying mitigation strategies. In the course of developing multiple options to meet strategic and military end states and objectives, JFCs and their planning staffs, as well as the larger JPEC, identify and communicate shortfalls in DOD's ability to resource, execute, and sustain the military operations contained in the plan as well as the necessary actions to reduce or mitigate risk. JFCs communicate risk to senior leadership. Risk is rarely eliminated, but through planning, preparation, and constant assessment, risk can be mitigated and managed.

f. Joint operation planning and planning for a campaign are not separate planning types or processes. Joint operation planning encompasses planning for any type of joint operation, such as small-scale, short-duration strike or raid; an operation that typically does not involve combat such as nation assistance; and large-scale, long-duration campaigns. Functional components (air, land, maritime, and special operations [SO]), Service components, and CSAs do not plan campaigns, but instead plan and conduct subordinate and supporting operations to campaign plans.

I-20   JP 1

Theory and Foundations

g. Joint operation planning requires the support of a wide array of staff expertise (personnel, intelligence, operations, logistics, communications, etc.) to provide JFCs with a thoughtful and coordinated product. Joint operation planning should be synchronized with national planning, so that interagency inputs are used in conjunction with military plans to reach strategic and military end states.

For more information, refer to JP 5-0, Joint Operation Planning; *JP 1-0,* Joint Personnel Support; *JP 2-0,* Joint Intelligence; *JP 3-0,* Joint Operations; *JP 4-0,* Joint Logistics; *and JP 6-0,* Joint Communications System.

13. Law of War

It is DOD policy that the Armed Forces of the United States will adhere to the law of war, often called the law of armed conflict, during all military operations. The law of war is the body of law that regulates both the legal and customary justifications for utilizing force and the conduct of armed hostilities; it is binding on the US and its individual citizens. It includes treaties and international agreements to which the US is a party, as well as applicable customary international law. It specifically applies to all cases of declared war or any other armed conflict involving the US; by policy, the principles and spirit of the law of war apply to all other military operations short of armed conflict. CCDRs must be particularly aware of the status of any conflict and the characterization of adversarial combatants and noncombatants (e.g., medical and chaplain personnel).

For further guidance on the law of war, refer to Chairman of the Joint Chiefs of Staff Instruction (CJCSI) 5810.01, Implementation of the DOD Law of War Program, *and JP 1-04,* Legal Support to Military Operations.

Chapter I

Intentionally Blank

CHAPTER II
DOCTRINE GOVERNING UNIFIED DIRECTION OF ARMED FORCES

> *"An army is a collection of armed men obliged to obey one man. Every change in the rules which impairs the principle weakens the army."*
>
> **William T. Sherman**
> **General of the Army, 1879**

1. National Strategic Direction

a. **National strategic direction** is governed by the Constitution, US law, USG policy regarding internationally recognized law, and the national interest as represented by national security policy. This direction leads to unified action which results in unity of effort to achieve national goals. At the strategic level, unity of effort requires coordination among government departments and agencies within the executive branch, between the executive and legislative branches, with nongovernmental organizations (NGOs), intergovernmental organizations (IGOs), the private sector, and among nations in alliance or coalition, and during bilateral or multilateral engagement. National policy and planning documents generally provide national strategic direction. The President and SecDef, through CJCS, provide direction for Service Chiefs, Military Department Secretaries, CCDRs, and CSA directors to:

(1) Provide clearly defined and achievable national strategic objectives.

(2) Provide timely strategic direction.

(3) Prepare Active Component (AC) and Reserve Component (RC) forces for combat.

(4) Focus DOD intelligence systems and efforts on the operational environment.

(5) Integrate DOD, partner nations, and/or other government departments and agencies into planning and subsequent operations.

(6) Maintain all required support assets in a high state of readiness.

(7) Deploy forces and sustaining capabilities that are ready to support the JFC's concept of operations (CONOPS).

Refer to JP 3-0, Joint Operations, *and JP 5-0,* Joint Operation Planning, *for more information on specific policy and planning documents related to national strategic direction.*

b. **Responsibilities for national strategic direction** as established by the Constitution and US law and practice are as follows:

II-1

Chapter II

(1) **The President of the United States** is responsible to the American people for national strategic direction.

(a) When the US undertakes military operations, the Armed Forces of the United States are only one component of a national-level effort involving all instruments of national power. Instilling unity of effort at the national level is necessarily a cooperative endeavor involving a number of USG departments and agencies. In certain operations, agencies of states, localities, or foreign countries may also be involved. The President establishes guidelines for civil-military integration and disseminates decisions and monitors execution through the NSC.

(b) Complex operations may require a high order of civil-military integration. Presidential directives guide participation by all US civilian and military agencies in such operations. Military leaders should work with the members of the national security team in the most skilled, tactful, and persistent ways to promote unity of effort. Operations of departments or agencies representing the diplomatic, economic, and informational instruments of national power are not under command of the Armed Forces of the United States or of any specific JFC. In US domestic situations, another department such as the Department of Homeland Security (DHS) may assume overall control of interagency coordination including military elements. Abroad, the chief of mission, supported by the country team, is normally in control.

(2) **SecDef** is responsible to the President for creating, supporting, and employing military capabilities. SecDef is the link between the President and the CCDRs, and provides direction and control of the CCDRs as they conduct military activities and operations. SecDef provides authoritative direction and control over the Services through the Secretaries of the Military Departments. SecDef exercises control of and authority over those forces not specifically assigned to the combatant commands (CCMDs) and administers this authority through the Military Departments, the Service Chiefs, and applicable chains of command. The Secretaries of the Military Departments organize, train, and equip forces and provide for the administration and support of forces within their department, including those assigned or attached to the CCDRs.

(3) **The CJCS** is the principal military advisor to the President, the NSC, and SecDef and functions under the authority, direction, and control of the President and SecDef. The CJCS assists the President and SecDef in providing for the strategic direction of the Armed Forces. Communications between the President or SecDef and the CCDRs are normally transmitted through the CJCS.

(4) **CCDRs** exercise combatant command (command authority) (COCOM) over assigned forces and are responsible to the President and SecDef for the preparedness of their commands and performance of assigned missions. GCCs have responsibility for a geographic AOR assigned through the Unified Command Plan (UCP). The UCP establishes CCMD missions and responsibilities, delineates the general geographical AOR for GCCs, and provides the framework used to assign forces. Functional combatant commanders (FCCs) have transregional responsibilities for assigned functions and support

II-2   JP 1

(or can be supported by) GCCs or may conduct missions assigned by the UCP independently.

(5) The Chief, National Guard Bureau (CNGB), is a principal advisor to SecDef through the CJCS on matters involving non-federalized National Guard forces and through other DOD officials as determined by SecDef. In domestic US situations, National Guard forces are a unique multi-status component with roles and responsibilities defined by federal and state law.

(6) In a foreign country, **the US chief of mission** is responsible to the President for directing, coordinating, and supervising all USG elements in the HN, except those under the command of a CCDR. GCCs are responsible for coordinating with chiefs of mission in their geographic AOR (as necessary) and for negotiating memoranda of agreement (MOAs) with the chiefs of mission in designated countries to support military operations. Force protection is an example of a military operation/requirement where an MOA would enhance coordination and integration.

2. Strategic Guidance and Responsibilities

a. **Military Planning.** Military planning consists of **joint strategic planning** with its three subsets: **security cooperation planning, force planning,** and **joint operation planning.** Regarding force planning for the future, DOD conducts capabilities-based planning (CBP). The essence of CBP is to identify capabilities that adversaries could employ against the US or a multinational opponent and to defend themselves; identify capabilities, US and multinational, that could be available to the joint or combined force to counter/defeat the adversary; and then identify and evaluate possible outcomes (voids or opportunities), rather than forecasting (allocating) forces against specific threat scenarios. Integral to a capabilities-based approach are joint capability areas (JCAs), DOD's capability management language and framework. JCAs are collections of like DOD capabilities functionally grouped to support capability analysis, strategy development, investment, decision making, capability portfolio management, and capabilities-based force development. They link the strategies for developing, managing, and employing the force by providing an organizing construct that facilitates collaboration among the many related DOD activities and processes. As the specific capabilities for given JCAs mature, they are vetted and verified through best practices based on extant capabilities and, when appropriate, incorporated into joint doctrine.

b. **National Planning Documents**

(1) The national security strategy (NSS), signed by the President, addresses the tasks that, as a Nation, are necessary to provide enduring security for the American people and shape the global environment. It provides a broad strategic context for employing military capabilities in concert with other instruments of national power. In the ends, ways, and means construct, the NSS provides the ends.

(2) The national defense strategy (NDS), signed by SecDef, outlines DOD's approach to implementing the President's NSS. The NDS supports the NSS by establishing

Chapter II

a set of overarching defense objectives that guide DOD's security activities and provide direction for the NMS. The NDS objectives serve as links between military activities and those of other USG departments and agencies in pursuit of national goals. This document provides the ways in the ends, ways, and means construct.

(3) The NMS, signed by CJCS, supports the aims of the NSS and implements the NDS. It describes the Armed Forces' plan to achieve military objectives in the near term and provides a vision for maintaining a force capable of meeting future challenges. It also provides focus for military activities by defining a set of interrelated military objectives and joint operating concepts from which the CCDRs and Service Chiefs identify desired capabilities and against which the CJCS assesses risk. This provides the final piece of the ends, ways, and means construct—the means.

(4) Guidance for Employment of the Force. The GEF merges Contingency Planning Guidance and Security Cooperation Guidance into one document that provides comprehensive, near-term planning guidance. The GEF provides Presidential and SecDef politico-military guidance. The President approves the contingency planning guidance contained in the GEF and approves SecDef's issuance of the GEF. The GEF is guided by the UCP and NDS and forms the basis for strategic policy guidance, campaign plans, and the Joint Strategic Capabilities Plan (JSCP).

(5) Joint Strategic Capabilities Plan. The JSCP provides guidance to CCDRs, Service Chiefs, CSA directors, applicable DOD agencies, DOD field activity directors, and the CNGB to accomplish tasks and missions based on near-term military capabilities. The JSCP is signed by the CJCS and implements campaign, contingency, and posture planning guidance reflected in the GEF.

(6) Further, the GEF and JSCP provide CCDRs with specific planning guidance for preparation of their theater campaign plans (TCPs), global campaign plans (GCPs), subordinate campaign plans, and contingency plans. Figure II-1 illustrates the various strategic guidance sources described below in the context of national strategic direction.

(7) The National Strategy for Homeland Security, also signed by the President, provides national direction to secure the homeland through a comprehensive framework for organizing the efforts of federal, state, local, tribal, and private organizations whose primary functions are often unrelated to national security.

(8) The National Response Framework developed by DHS establishes a comprehensive, national-level, all-hazards, all-discipline approach to domestic incident management. It covers the full range of complex and constantly changing requirements in anticipation of, or in response to, threats or acts of terrorism, major disasters, and other emergencies. DOD develops or revises its plans to align with this framework and effectively and efficiently employ the joint force.

II-4   JP 1

Doctrine Governing Unified Direction of Armed Forces

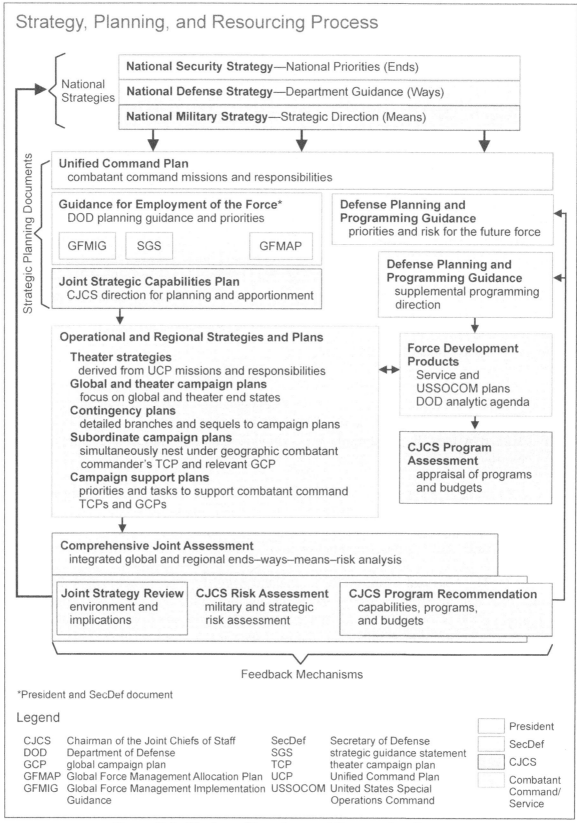

Figure II-1. Strategy, Planning, and Resourcing Process

Chapter II

(a) A TCP is based on planning guidance provided by the GEF and JSCP. A TCP operationalizes CCDR functional and theater strategies. Campaign plans focus on the command's steady-state (Phase 0) activities, which include ongoing operations, security cooperation, and other shaping or preventive activities for the next 5 years. It should include measurable and achievable objectives that contribute to the strategic end states in the GEF. Contingency plans for responding to crisis scenarios should be treated as branch plans to the campaign plan. For planning purposes, GCCs use assigned forces, those rotationally deployed into the AOR, and those forces that historically have been deployed for engagement activities. Each GCC's TCP and FCC's GCP are sent to CJCS for review and integration into the global family of TCPs.

(b) Campaign Support Plans. Supporting CCMDs, Services, and DOD agencies routinely conduct security cooperation activities within a GCC's AOR or involving foreign nationals from countries within an AOR. Services and select DOD agencies will coordinate and provide their security cooperation strategies to the supported GCC.

<u>1</u>. Campaign support plans will balance competing CCMD demands for limited global resources.

<u>2</u>. Campaign support plans or their update memoranda are submitted to the CJCS and Under Secretary of Defense for Policy for review annually and are shared with the GCCs.

c. **Role of the Geographic Combatant Commanders**

(1) GCCs are the vital link between those who determine national security policy and strategy and the military forces or subordinate JFCs that conduct military operations within their AORs. GCCs are responsible for a large geographical area and for effective coordination of operations within that area. Directives flow from the President and SecDef through CJCS to the GCCs, who plan and conduct the operations that achieve national or multinational strategic objectives. GCCs provide guidance and direction through strategic estimates, command strategies, and plans and orders for the employment of military force. As military force may not achieve national objectives, it must be coordinated, synchronized, and if appropriate, integrated with other USG departments and agencies, IGOs, NGOs, multinational forces (MNFs), and elements of the private sector.

(2) Using their strategic estimate(s) and strategic options, GCCs develop strategies that translate national and multinational direction into strategic concepts or courses of action (COAs) to meet strategic and joint operation planning requirements. GCCs' plans provide strategic direction; assign missions, tasks, forces, and resources; designate objectives; provide authoritative direction; promulgate ROE and rules for the use of force (RUF); establish constraints and restraints (military limitations); and define policies and CONOPS to be integrated into subordinate or supporting plans. GCCs also exercise directive authority for logistics over assigned forces and authority for force protection over all DOD personnel (including their dependents) assigned, attached,

transiting through, or training in the GCC's AOR. The exception is for those for whom a chief of mission retains security responsibility.

d. **Functional Combatant Commanders.** FCCs provide support to and may be supported by GCCs and other FCCs as directed by higher authority. FCCs are responsible for a large functional area requiring single responsibility for effective coordination of the operations therein. These responsibilities are normally global in nature. The President and SecDef direct what specific support and to whom such support will be provided. When an FCC is the supported commander and operating within GCCs' AORs, close coordination and communication between them is paramount.

e. **Service Chiefs and Commander, United States Special Operations Command (CDRUSSOCOM).** The Service Chiefs and CDRUSSOCOM (in areas unique to SO) under authority established in Title 10, USC, among other tasks, organize, train, and equip AC and RC forces, DOD civilian personnel, contractor personnel, and selected HN personnel. The AC and RC are fully integrated partners in executing US military strategy, to include HD and defense support of civil authorities (DSCA) operations. The RC provides operational capabilities and strategic depth to meet US requirements worldwide. The RC provides operational forces that can be used on a regular basis while maintaining strategic depth in the event of mid- to large-scale contingencies or other unanticipated national crises. Unpredictable crises call for trained and ready forces that are either forward deployed or are rapidly and globally deployable. These forces should be initially self-sufficient and must possess the capabilities needed to effectively act in the US national interest or signal US resolve prior to conflict. Such forces are usually drawn from the **active force structure** and normally are tailored and integrated into joint organizations that capitalize on the unique and complementary capabilities of the Services and United States Special Operations Command (USSOCOM).

f. **United States Coast Guard (USCG).** The Commandant of the Coast Guard is responsible for organizing, training, and equipping Service forces under Titles 10 and 14, USC. The Commandant may provide forces to GCCs to perform activities for which those forces are uniquely suited. Under Title 14, USC, the USCG is assigned to DHS for homeland security (HS). In addition, the Commandant is responsible for the coordination and conduct of maritime law enforcement and security operations under civil authorities for HS in the US maritime domain. DOD forces may act in direct support of USCG commanders. The USCG has authority to make inquiries, examinations, inspections, searches, seizures, and arrests upon the high seas and waters over which the US has jurisdiction. It is the only military Service not constrained by the Posse Comitatus Act or its extension by DOD directive.

g. **DOD Agencies.** DOD agencies are organizations established by SecDef under Title 10, USC, to perform a supply or service activity common to more than one Military Department. There are 16 DOD agencies including Defense Intelligence Agency (DIA), Defense Logistics Agency (DLA), Missile Defense Agency, and Defense Threat Reduction Agency (DTRA), among others.

Chapter II

3. Unified Action

a. **Unified action** synchronizes, coordinates, and/or integrates joint, single-Service, and multinational operations with the operations of other USG departments and agencies, NGOs, IGOs (e.g., the United Nations [UN]), and the private sector to achieve **unity of effort** (see Figure II-2). Unity of command within the military instrument of national power **supports the national strategic direction** through close coordination with the other instruments of national power.

b. **Success often depends on unified actions.** The CJCS and all CCDRs are in pivotal positions to facilitate the planning and conduct of unified actions in accordance with the guidance and direction received from the President and SecDef in coordination with other authorities (i.e., multinational leadership).

c. Unity of command in the Armed Forces of the United States **starts with national strategic direction.** For US military operations, unity of command is accomplished by establishing a joint force, assigning a mission or objective(s) to the designated JFC, establishing command relationships, assigning and/or attaching appropriate forces to the

Figure II-2. Unified Action

joint force, and empowering the JFC with sufficient authority over the forces to accomplish the assigned mission.

4. Roles and Functions

The terms "roles and functions" should not be used interchangeably, as the distinctions between them are important.

a. Roles are the broad and enduring purposes for which the Services and the CCMDs were established in law.

b. Functions are the appropriate assigned duties, responsibilities, missions, or tasks of an individual, office, or organization. As defined in the National Security Act of 1947, as amended, the term "function" includes functions, powers, and duties (Title 50, USC, Section 410[a]).

For further information on functions, refer to Chapter I, "Theory and Foundations," Paragraph 7, "Task, Function, and Mission," and "Chapter III, "Functions of the Department of Defense and Its Major Components."

5. Chain of Command

The President and SecDef exercise authority, direction, and control of the Armed Forces through two distinct branches of the chain of C2 (see Figure II-3). One branch runs from the President, through SecDef, to the CCDRs for missions and forces assigned to their commands. For purposes other than the operational direction of the CCMDs, the chain of command runs from the President to SecDef to the Secretaries of the Military Departments and, as prescribed by the Secretaries, to the commanders of Military Service forces. The Military Departments, organized separately, operate under the authority, direction, and control of the Secretary of that Military Department. The Secretaries of the Military Departments may exercise administrative control (ADCON) over Service forces through their respective Service Chiefs and Service commanders. The Service Chiefs, except as otherwise prescribed by law, perform their duties under the authority, direction, and control of the Secretaries of the respective Military Departments to whom they are directly responsible.

a. The CCDRs exercise COCOM over assigned forces and are directly responsible to the President and SecDef for the performance of assigned missions and the preparedness of their commands. CCDRs prescribe the chain of command within their CCMDs and designate the appropriate command authority to be exercised by subordinate commanders.

b. The Secretaries of the Military Departments operate under the authority, direction, and control of SecDef. This branch of the chain of command is responsible for ADCON over all military forces within the respective Service not assigned to CCDRs (i.e., those defined in the Global Force Management Implementation Guidance [GFMIG] as "unassigned forces"). This branch is separate and distinct from the branch of the chain of command that exists within a CCMD.

Chapter II

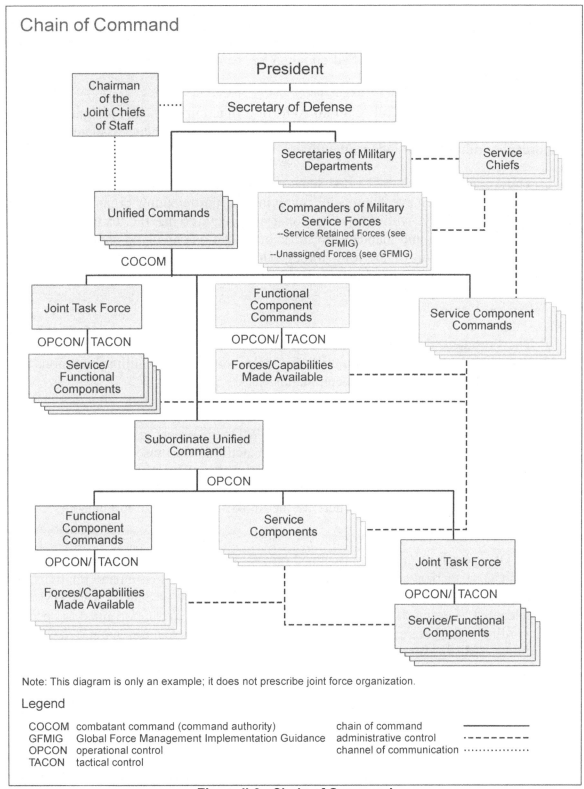

Figure II-3. Chain of Command

6. Unified Command Plan

a. The President, through the UCP, establishes CCMDs. Commanders of unified CCMDs may establish subordinate unified commands when so authorized by SecDef. SecDef, CCDR, a subordinate unified commander, or an existing JTF commander may establish JTFs.

b. CCDRs have responsibility for an AOR or a function (e.g., SO) assigned through the UCP. FCCs support (or can be supported by) GCCs or may conduct assigned missions in accordance with the UCP independently.

c. The Armed Forces of the United States are most effective when employed as a joint force. This "comprehensive approach" involving all participating organizations, both military and nonmilitary, within an operational area requires the JFC to understand the capabilities, limitations, and mandates of those organizations involved and to effectively communicate the mission of the joint force. The basic doctrinal foundations for joint functions at all levels are outlined in this chapter.

7. Combatant Commands

a. The President, through SecDef and with the advice and assistance of the CJCS, establishes combatant (unified) commands for the performance of military missions and prescribes the force structure of such commands.

b. The CJCS assists the President and SecDef in performing their command functions. The CJCS transmits to the commanders of the CCMDs the orders given by the President or SecDef and, as directed by SecDef, oversees the activities of those commands. Orders issued by the President or SecDef normally are conveyed by the CJCS under the authority and direction of SecDef. Reports from CCDRs normally will be submitted through CJCS, who forwards them to SecDef and acts as the spokesman for the commanders of the CCMDs.

c. CCDRs exercise COCOM of assigned forces. The CCDR may delegate operational control (OPCON), tactical control (TACON), or establish support command relationships of assigned forces. Unless otherwise directed by the President or SecDef, COCOM may not be delegated. During deliberate planning, generic forces are apportioned to specific plans according to Global Force Management procedures. This requires supported CCDRs to coordinate with the supporting CCDRs and Services on required capabilities during planning and on mission criteria for specific units once they have been allocated.

8. Military Departments, Services, Forces, Combat Support Agencies, and National Guard Bureau

a. The authority vested in the Secretaries of the Military Departments in the performance of their role to organize, train, equip, and provide forces runs from the President through SecDef to the Secretaries. Then, to the degree established by the Secretaries or specified in law, this authority runs through the Service Chiefs to the Service component commanders assigned to the CCDRs and to the commanders of forces not

Chapter II

assigned to the CCDRs. ADCON provides for the preparation of military forces and their administration and support, unless such responsibilities are specifically assigned by SecDef to another DOD component.

b. The Secretaries of the Military Departments are responsible for the administration and support of Service forces. They fulfill their responsibilities by exercising ADCON through the Service Chiefs. Service Chiefs have ADCON for all forces of their Service. The responsibilities and authority exercised by the Secretaries of the Military Departments are subject by law to the authority provided to the CCDRs in their exercise of COCOM.

c. Each of the Secretaries of the Military Departments, coordinating as appropriate with the other Military Department Secretaries and with the CCDRs, has the responsibility for organizing, training, equipping, and providing forces to fulfill specific roles and for administering and supporting these forces. The Secretaries also perform a role as a force provider of Service retained forces until they are deployed to CCMDs. When addressing similar issues regarding National Guard forces, coordination with the National Guard Bureau (NGB) is essential.

d. Commanders of Service forces are responsible to Secretaries of the Military Departments through their respective Service Chiefs for the administration, training, and readiness of their unit(s). Commanders of forces assigned to the CCMDs are under the authority, direction, and control of (and are responsible to) their CCDR to carry out assigned operational missions, joint training and exercises, and logistics.

e. The USCG is a military Service and a branch of the US Armed Forces at all times. However, it is established separately by law as a Service in DHS, except when transferred to the Department of the Navy (DON) during time of war or when the President so directs. Authorities vested in the USCG under Title 10, USC, as an armed Service and Title 14, USC, as a federal maritime safety and law enforcement agency remain in effect at all times, including when USCG forces are operating within DOD/DON chain of command. USCG commanders and forces may be attached to JFCs in performance of any activity for which they are qualified. Coast Guard units routinely serve alongside Navy counterparts operating within a naval task organization in support of a maritime component commander.

f. The NGB is a joint activity of DOD. The NGB performs certain military Service-specific functions and unique functions on matters involving non-federalized National Guard forces. The NGB is responsible for ensuring that units and members of the Army National Guard and the Air National Guard are trained by the states to provide trained and equipped units to fulfill assigned missions in federal and non-federal statuses.

g. In addition to the Services above, a number of DOD agencies provide combat support or combat service support to joint forces and are designated as CSAs. CSAs, established under SecDef authority under Title 10, USC, Section 193, and Department of Defense Directive (DODD) 3000.06, *Combat Support Agencies,* are the DIA, National Geospatial-Intelligence Agency (NGA), Defense Information Systems Agency (DISA), DLA, Defense Contract Management Agency (DCMA), DTRA, and National Security

Agency (NSA). These CSAs provide CCDRs specialized support and operate in a supporting role. The CSA directors are accountable to SecDef.

9. Relationship Among Combatant Commanders, Military Department Secretaries, Service Chiefs, and Forces

a. **Continuous Coordination.** The Services and USSOCOM (in areas unique to SO) share the division of responsibility for developing military capabilities for the CCMDs. All components of DOD are charged to coordinate on matters of common or overlapping responsibility. The Joint Staff, Services, and USSOCOM headquarters play a critical role in ensuring that CCDRs' concerns and comments are included or advocated during the coordination.

b. **Interoperability.** Unified action demands maximum interoperability. The forces, units, and systems of all Services must operate together effectively, in part through interoperability. This includes joint force development; use of joint doctrine; the development and use of joint plans and orders; and the development and use of joint and/or interoperable communications and information systems. It also includes conducting joint training and exercises. It concludes with a materiel development and fielding process that provides materiel that is fully compatible with and complementary to systems of all Services. A key to successful interoperability is to ensure that planning processes are joint from their inception. Those responsible for systems and programs intended for joint use will establish working groups that fully represent the services and functions affected. CCDRs will ensure maximum interoperability and identify interoperability issues to the CJCS, who has overall responsibility for the joint interoperability program. Other government departments and agencies, IGOs, and NGOs should be invited to participate in joint training and exercises whenever possible.

10. Interagency Coordination

a. **General**

(1) Interagency coordination is the cooperation and communication that occurs between departments and agencies of the USG to accomplish an objective. Similarly, in the context of DOD involvement, coordination refers to coordination between elements of DOD and IGOs or NGOs to achieve objectives.

(2) CCDRs and subordinate JFCs must consider the potential requirements for interagency, IGO, and NGO coordination as a part of their activities within and outside of their operational areas. Military operations must be coordinated, integrated, and deconflicted with the activities of interorganizational partners, including various HN agencies within and en route to and from the operational area. Sometimes the JFC draws on the capabilities of other organizations, provides capabilities to other organizations, and sometimes the JFC merely deconflicts activities with those of others. These same organizations may be involved during all phases of an operation including pre- and post-operation activities. Roles and relationships among USG departments and agencies, state, tribal, and local governments, must be clearly understood. Interagency coordination forges

Chapter II

the vital link between the military and the diplomatic, informational, and economic instruments of national power. Successful interorganizational coordination helps enable the USG to build international and domestic support, conserve resources, and conduct coherent operations that efficiently achieve shared goals.

For more information on interagency coordination, see JP 3-08, Interorganizational Coordination During Joint Operations.

b. **Interagency Unity of Effort**

(1) **Achieving Unity of Effort.** Some of the techniques, procedures, and systems of military C2 can facilitate unity of effort if they are adjusted to the dynamic world of interagency coordination and different organizational cultures. Unity of effort can only be achieved through close, continuous interagency and interdepartmental coordination and cooperation, which are necessary to overcome discord, inadequate structure and procedures, incompatible communications, cultural differences, and bureaucratic and personnel limitations.

(2) **Unity of Effort Guidance.** Within the USG, the NSS guides the development, integration, and coordination of all the instruments of national power to accomplish national objectives. The NSC is the principal policymaking forum responsible for the strategic-level implementation of the NSS. The NSC system is a process to coordinate executive departments and agencies in the effective development and implementation of those national security policies. This coordination sets the stage for strategic guidance provided to the CCMDs, Services, and various DOD agencies and forms the foundation for operational and tactical level guidance.

(3) **National Security Council.** The NSC is the principal forum for consideration of national security policy issues requiring Presidential determination. The NSC advises and assists the President in integrating all aspects of national security policy—domestic, foreign, military, intelligence, and economic (in conjunction with the National Economic Council). Along with its subordinate committees, the NSC is the principal means for coordinating, developing, and implementing national security policy. The statutory members of the NSC are the President, Vice President, Secretary of State, and SecDef. The CJCS is the council's statutory military advisor and the Director of National Intelligence is the council's statutory intelligence advisor. Officials from the Office of the Secretary of Defense (OSD) represent SecDef in NSC interagency groups. Similarly, the CJCS, assisted by the Joint Staff, represents the CCDRs for interagency matters in the NSC system. Other senior officials are invited to attend NSC meetings, as appropriate. Subcommittees of the NSC include:

(a) **NSC Principals Committee (PC).** NSC/PC is the senior (cabinet-level) interagency forum for considering policy issues affecting national security.

(b) **NSC Deputies Committee (DC).** NSC/DC is the senior subcabinet interagency forum for considering policy issues affecting national security. The NSC/DC prescribes and reviews the work of the NSC interagency groups, helps to ensure that issues

II-14   JP 1

brought before the NSC/PC or the NSC have been properly analyzed and prepared for decision, and oversees day-to-day crisis management and prevention.

(c) **NSC Interagency Policy Committees (IPCs).** The main day-to-day forum for interagency coordination of national security policy, NSC/IPCs manage the development and implementation of national security policies by multiple departments and agencies of the USG. NSC/IPCs provide policy analysis for the more senior committees of the NSC system to consider and ensure timely responses to Presidential decisions. IPCs are grouped as either regional or functional.

(d) **Regional NSC/IPCs.** Regional NSC/IPCs may be established and chaired by the appropriate Under or Assistant Secretary of State with responsibility for regional issues (e.g., Europe and Eurasia, Western Hemisphere, East Asia).

(e) **Functional NSC/IPCs.** NSC/IPCs, each chaired at the Under or Assistant Secretary level within the agency indicated, have been established for 15 various functional topics. Some of them include: democracy, human rights, and international operations; counterterrorism and national preparedness; defense strategy, force structure, and planning (DOD); arms control; proliferation, counterproliferation, and HD; intelligence and counterintelligence; international organized crime; contingency planning; space; and international drug control.

(4) While the NSC serves as the principal forum for considering national security policy issues requiring Presidential determination, the HSC provides a parallel forum for considering unique HS matters, especially those concerning terrorism within the US.

For more information on the NSC, see CJCSI 5715.01, Joint Staff Participation in Interagency Affairs, *and National Security Presidential Directive-1,* Organization of the National Security Council System.

(5) **Homeland Security Council.** The HSC is responsible for advising and assisting the President with respect to all aspects of HS, and serves as the mechanism for ensuring coordination of HS-related activities of executive departments and agencies and effective development and implementation of HS policies. Other subcommittees of the HSC include:

(a) **HSC Principals Committee.** The HSC/PC is the senior (cabinet-level) interagency forum for HS issues.

(b) **HSC Deputies Committee.** The HSC/DC is the senior subcabinet interagency forum for consideration of policy issues affecting HS. The HSC/DC tasks and reviews the work of the HSC interagency groups and helps ensure that issues brought before the HSC/PC or HSC have been properly analyzed and prepared for action.

(c) **HSC Interagency Policy Committees.** The main forum for interagency coordination of HS policy, HSC-IPCs coordinate the development and implementation of HS policies by multiple departments and agencies throughout the USG and coordinate those policies with state and local government. HSC-IPCs provide policy analysis for

Chapter II

consideration by the more senior committees of the HSC system and ensure timely responses to Presidential decisions. There are 11 HSC-IPCs established for functional areas such as: detection, surveillance, and intelligence (intelligence and detection); plans, training, exercises, and evaluation (policy and plans); WMD consequence management (response and recovery); key asset, territorial waters and airspace, and security (protection and prevention); and domestic threat response and incident management (response and recovery).

For more information on the HSC, see CJCSI 5715.01, Joint Staff Participation in Interagency Affairs, *and Homeland Security Presidential Directive-1,* Organization and Operations of the Homeland Security Council.

c. **Interagency Coordination and Integration**

(1) The guidelines for interagency coordination ensure that all participating departments and agencies under appropriate authority focus their efforts on national objectives. The Armed Forces of the United States have unique capabilities to offer the interagency community. These include established military-to-military domestic and international contacts, resources (e.g., logistics) not available to nonmilitary agencies, trained civil affairs (CA) personnel and their assets, and responsiveness based on military training and readiness. Additional unique military capabilities include C2 resources supported by worldwide communications and intelligence, surveillance, and reconnaissance (ISR) infrastructures; cyberspace capabilities; robust organizational and planning processes; training support for large numbers of individuals on myriad skills; and air, land, and maritime mobility support for intertheater or intratheater requirements.

(2) **Interorganizational Coordination in Foreign Areas**

(a) Interorganizational coordination in foreign areas may involve the exercise of USG policy regarding internationally recognized law; preexisting bilateral and multilateral military relationships, agreements, and arrangements managed by US embassies; treaties involving US defense interests, implementation of CCMD theater security cooperation activities; and initiatives concerning technology transfer or armaments cooperation and control, foreign humanitarian assistance, peace operations, or other contingencies.

(b) At the national level, DOS leads the effort to support interagency coordination overseas, forming task-oriented groups or employing the NSC system to organize the effort.

(c) The formal US interagency structure in foreign countries operates under the lead of the US chief of mission, normally an ambassador, and the country team and may include US embassy PA and cultural affairs representation. The chief of mission is ordinarily the lead for interagency coordination abroad that is essentially nonmilitary in nature but requires military participation, with representation and control of the military operations provided by the JFC.

II-16   JP 1

Doctrine Governing Unified Direction of Armed Forces

(d) Within an AOR, the GCC is responsible for planning and implementing military strategies and operations and interorganizational coordination. Coordination required outside the geographic region may be supported by groups within the NSC system or individual USG departments and agencies, with lead for such coordination falling to the CCMD or the USG department or agency, depending on the circumstances. In some operations, a special representative of the President or special envoy of the UN Secretary-General may be involved.

(3) **Domestic Interagency Coordination**

(a) For HS-related interagency coordination that may require military participation in countering domestic terrorism and other support tasks, DHS has the lead. For HD interagency coordination, DOD will have the lead. DHS is the primary forum for coordinating executive branch efforts to detect, prepare for, prevent, protect against, respond to, and recover from terrorist attacks within the US.

(b) In domestic situations, US law and policy limit the scope and nature of military actions. SecDef retains the authority to approve use of DOD resources for assistance to civil authorities. For DSCA operations within the US, the Joint Staff Joint Director of Military Support (JDOMS) validates requests for assistance, determines what DOD capabilities are available to fulfill the request, and coordinates SecDef approval to use DOD forces. JDOMS will normally allocate Title 10, USC, forces to United States Northern Command (USNORTHCOM) for operations approved by SecDef. The National Guard has unique roles in domestic operations. The National Guard in either state active duty or Title 32, USC, status (not in federal service) is likely to be the first military force to provide support for an incident.

(c) Per the Posse Comitatus Act and DOD regulations, the US is generally prohibited from employing Title 10, USC, DOD forces to provide direct military involvement to enforce the law of the US unless expressly authorized by the Constitution or Congress. For example, the President, as Commander in Chief under the Insurrection Act, may use the military in cases of civil disturbance and to protect USG functions and property. It is important to note that use of military forces in the defense of the US is not support to civilian law enforcement and is not subject to the prohibitions of the Posse Comitatus Act.

(d) In addition to coordinating with USG departments and agencies, other domestic participants may be involved, to include state, local, and/or tribal government organizations as well as the types of NGOs and IGOs that operate domestically and/or internationally.

For more information on HS, HD, DSCA, and associated interagency coordination activities in support of these missions, see the National Strategy for Homeland Security, National Response Framework, DOD Strategy for Homeland Defense and Civil Support, *JP 3-27,* Homeland Defense, *and JP 3-28,* Civil Support.

(4) **Command Relationships**

II-17

Chapter II

(a) Command relationships preserve the primacy of civil authorities in their spheres of responsibility while facilitating the full utilization of military forces as permitted by the Constitution, law, and directives of the President. Military commands provide assistance in consonance with these directives for activities conducted under the control of civil authorities.

(b) The relationship between NGOs, IGOs, and US military elements may be viewed as an associate or partnership relationship. These civilian organizations do not operate in military or governmental hierarchies and therefore do not have formal supporting or supported relationships with US military forces. However, an MOA or memorandum of understanding can outline agreed to relationships.

(5) **Organizing for Interagency Coordination**

(a) **Joint Interagency Coordination Group (JIACG).** When formed, a JIACG can provide the CCDR with an increased capability to collaborate with other USG civilian agencies and departments (see Figure II-4 for a notional JIACG structure). The JIACG, an element of a CCDR's staff, is an interagency staff group that establishes and enhances regular, timely, and collaborative working relationships between other governmental agencies' representatives (DOS, DHS, and others) and military operational planners at the CCMDs. JIACGs complement the interagency coordination that occurs at the national level through DOD and the NSC and HSC systems. JIACG members participate in deliberate and crisis action planning. They provide a conduit back to their parent organizations to help synchronize joint operations with the efforts of USG departments and agencies.

(b) A contingency and planning focused subgroup of the JIACG is the interagency planning cell. The interagency planning cell can be organized or tailored to operate 24/7 to assist in and support interagency planning and/or coordination in crisis and/or contingency situations. During such situations, an interagency planning cell will enable a coherent, efficient, and responsive planning and coordination effort through focused or targeted participation by interagency subject matter experts and dedicated agency representatives. An interagency planning cell should be activated to support a CCMD's campaign planning efforts, ensuring interagency issues are fully considered in mission analysis and COA development.

For more information on the JIACG and the interagency planning cell, see JP 3-08, Interorganizational Coordination During Joint Operations.

(6) **JTF Interagency Coordination**

(a) There are specific policies and procedures that guide JTF interagency coordination. The unique aspects of the interagency process require the JTF headquarters to be especially flexible, responsive, and cognizant of the capabilities of not only the JTF's components, but other agencies as well.

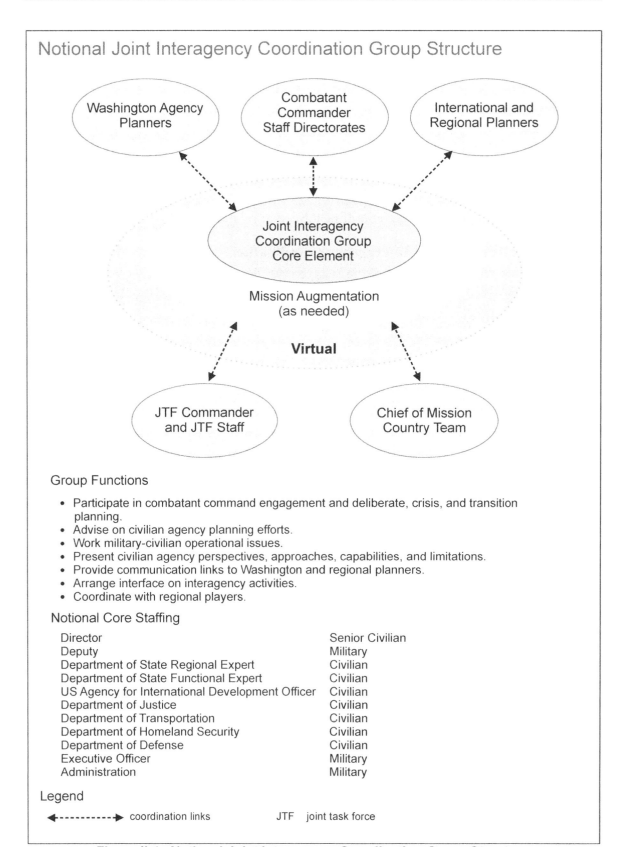

Figure II-4. Notional Joint Interagency Coordination Group Structure

Chapter II

(b) In contrast to the established command structure of a CCMD or JTF, NGOs and IGOs in the operational area may not have a defined structure for controlling activities. Upon identifying organizational or operational mismatches between organizations, the staff of the CCMD or JTF should coordinate points in the NGOs and IGOs at which liaison and coordinating mechanisms are appropriate.

(c) The civil-military operations center (CMOC) is composed of representatives from military, civilian, US, and multinational agencies involved in the operation (see Figure II-5). An effective CMOC contributes to meeting the objectives of all represented agencies in a cooperative and efficient manner. To best coordinate and collaborate military and civilian operations, the JTF should carefully consider where to locate the CMOC (i.e., proximity to the JTF command center).

For more information on the CMOC, see JP 3-08, Interorganizational Coordination During Joint Operations, *and JP 3-57,* Civil-Military Operations.

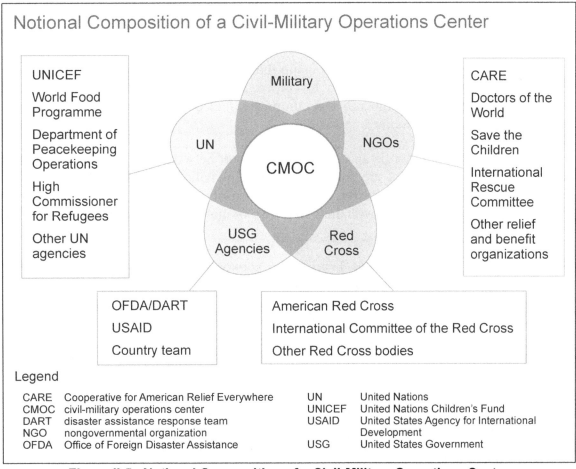

Figure II-5. Notional Composition of a Civil-Military Operations Center

11. Multinational Operations

a. General

(1) International partnerships continue to underpin unified efforts to address 21st century challenges. Shared principles, a common view of threats, and commitment to cooperation provide far greater security than the US could achieve independently. These partnerships must be nurtured and developed to ensure their relevance even as new challenges emerge. The ability of the US and its allies to work together to influence the global environment is fundamental to defeating 21st century threats. Wherever possible, the US works with or through other nations, enabling allied and partner capabilities to build their capacity and develop mechanisms to share the risks and responsibility of today's complex challenges.

(2) Operations conducted by forces of two or more nations are termed "multinational operations." Such operations are usually undertaken within the structure of a coalition or alliance. Other possible arrangements include supervision by an IGO such as the UN or the Organization for Security and Co-Operation in Europe. Other commonly used terms for multinational operations include allied, bilateral, or multilateral, as appropriate.

(a) An alliance is a relationship that results from a formal agreement (e.g., treaty) between two or more nations for broad, long-term objectives that further the common interests of the members. Operations conducted with units from two or more allies are referred to as combined operations.

(b) A coalition is an ad hoc arrangement between two or more nations for common action. Coalitions are formed by different nations with specific objectives, usually for a single occasion or for longer cooperation in a narrow sector of common interest. Operations conducted with units from two or more coalition members are referred to as coalition operations.

(3) Cultural, psychological, religious, economic, technological, informational, and political factors as well as transnational dangers all impact multinational operations. Many contingency plans to deter or counter threats are prepared within the context of a treaty or alliance framework. Sometimes they are developed in a less structured coalition framework, based on temporary agreements or arrangements. Much of the information and guidance provided for unified action and joint operations are applicable to multinational operations. However, differences in laws, doctrine, organization, weapons, equipment, terminology, culture, politics, religion, and language within alliances and coalitions must be considered. Normally, each alliance or coalition develops its own plans and orders to guide multinational action.

(4) No single command structure best fits the needs of all alliances and coalitions. Each coalition or alliance will create the structure that best meets the objectives, political realities, and constraints of the participating nations. Political considerations heavily influence the ultimate shape of the command structure. However, participating nations

Chapter II

should strive to achieve unity of effort for the operation to the maximum extent possible, with missions, tasks, responsibilities, and authorities clearly defined and understood by all participants. While command relationships are well defined in US doctrine, they are not necessarily part of the doctrinal lexicon of nations in an alliance or coalition.

b. **Multinational Unity of Effort.** Attaining unity of effort through unity of command for a multinational operation may not be politically feasible, but it should be a goal. There must be a common understanding among all national forces of the overall aim of the MNF and the plan for its attainment. A coordinated policy, particularly on such matters as multinational force commanders' (MNFCs') authority over national logistics (including infrastructure), ROE, fratricide prevention, and ISR is essential for unity of effort. While the tenets discussed below cannot guarantee success, ignoring them may lead to mission failure due to a lack of unity of effort.

(1) Respect. In assigning missions, the commander must consider that national honor and prestige may be as important to a contributing nation as combat capability. All partners must be included in the planning process and their opinions must be sought in mission assignment. Understanding, consideration, and acceptance of partner ideas are essential to effective communication across cultures, as are respect for each partner's culture, religion, customs, history, and values. Junior officers in command of small national contingents may be the senior representatives of their government within the MNFs and, as such, should be treated with special consideration beyond their US equivalent rank.

(2) Rapport. US commanders and staffs should establish rapport with their counterparts from partner countries, as well as the MNFC. This requires personal, direct relationships that only they can develop. Good rapport between leaders will improve teamwork among their staffs and subordinate commanders and overall unity of effort.

(3) Knowledge of Partners. US commanders and their staffs should have an understanding of each member of the MNF. Much time and effort is expended in learning about the enemy; a similar effort is required to understand the doctrine, capabilities, strategic goals, culture, religion, customs, history, and values of each partner. This will ensure the effective integration of multinational partners into the operation and enhance the synergistic effectiveness of the MNF.

(4) Patience. Effective partnerships take time and attention to develop. Diligent pursuit of a trusting, mutually beneficial relationship with multinational partners requires untiring, evenhanded patience. This is easier to accomplish within alliances but is equally necessary regarding prospective multinational partners.

(5) Coordination. Coordinated policy, particularly on such matters as MNFCs' authority over national logistics (including infrastructure) and ISR, is required. Coordinated planning for ROE, RUF, fratricide prevention, IO, communications, special weapons, source and employment of reserves, and timing of operations is essential for unity of effort. Actions to improve interoperability and the ability to share information need to be addressed early. This includes an emphasis on the uses of multinational doctrine and

II-22

JP 1

tactics, techniques, and procedures; development of ISR, C2 systems, and logistic architectures; multinational training and exercises; and establishment of liaison structures. Nations should exchange qualified liaison officers at the earliest opportunity to ensure mutual understanding and unity of effort.

c. **Multinational Organizational Structure**

(1) Organizational Structure. The basic structures for multinational operations fall into one of three types: integrated, lead nation, or parallel command.

(a) Integrated commands have representative members from the member nations in the command headquarters. Multinational commands organized under an integrated command help ensure the capabilities of member nations are represented and employed properly. A good example of this command structure is found in the North American Aerospace Defense Command (NORAD) where the commander is American, the deputy commander is Canadian, and each of the regional commands has a commander and deputy commander from a different nation. In addition, the NORAD staff is binational.

(b) Lead Nation Command Structure. A lead nation command structure exists when all member nations place their forces under the control of one nation. The lead nation command can be distinguished by a dominant lead nation command and staff arrangement with subordinate elements retaining strict national integrity.

(c) Parallel Command Structures. Under a parallel command structure, no single force commander is designated. The MNF leadership must develop a means for coordination among the participants to attain unity of effort. This can be accomplished through the use of coordination centers. Nonetheless, because of the absence of a single MNFC, the use of a parallel command structure should be avoided if possible.

(2) Regardless of how the MNF is organized operationally, each nation furnishing forces normally establishes a national component, often called a national command element, to ensure effective administration of its forces. The national component provides a means to administer and support the national forces, coordinate communication to the parent nation, tender national military views and recommendations directly to the MNFC, facilitate the assignment and reassignment of national forces to subordinate operational multinational organizations, and maintain personnel accountability. In an administrative role, these national components are similar to a Service component command at the unified CCMD level in a US joint organization. The logistic support element of this component is referred to as the national support element.

d. **Command and Control of US Forces in Multinational Operations.** Although nations will often participate in multinational operations, they rarely, if ever, relinquish national command of their forces. As such, forces participating in a multinational operation will always have at least two distinct chains of command: a national chain of command and a multinational chain of command (see Figure II-6).

Chapter II

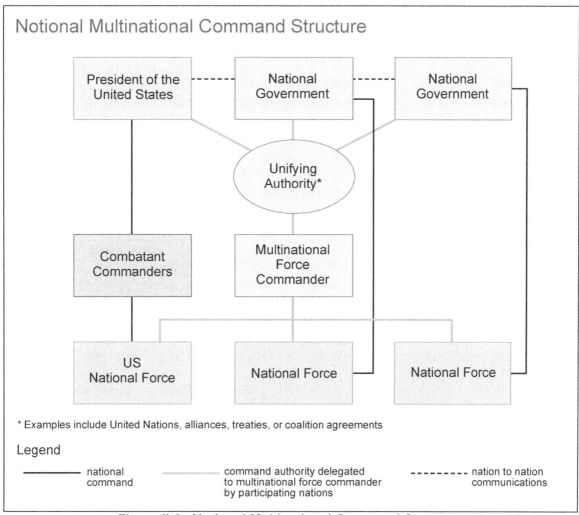

Figure II-6. Notional Multinational Command Structure

(1) National Command. The President retains and cannot relinquish national command authority over US forces. National command includes the authority and responsibility for organizing, directing, coordinating, controlling, planning employment, and protecting military forces. The President also has the authority to terminate US participation in multinational operations at any time.

(2) Multinational Command. Command authority for an MNFC is normally negotiated between the participating nations and can vary from nation to nation. Command authority will be specified in the implementing agreements and may include aspects of OPCON and/or TACON, as well as support relationships and coordinating authority. A clear and common understanding of what authorities are specified in the implementing agreement is essential to operations. This is particularly important when similar terms have different meanings to the various participants. For example, both the US and North Atlantic Treaty Organization (NATO) use the term operational control and the acronym OPCON, but the authorities of US OPCON are more encompassing than the authorities of NATO OPCON. Commanders and forces must be aware of which authorities are specified.

(a) Operational Control. While the President cannot relinquish command authority, in some multinational environments it might be prudent or advantageous to place appropriate US forces under the OPCON of an MNFC to achieve specified military objectives. In making this decision, the President carefully considers such factors as mission, size of the proposed US force, risks involved, anticipated duration, and ROE. Normally, OPCON of US forces is assigned only for a specific timeframe or mission and includes the authority to assign tasks to US forces already deployed by the President and to US units led by US officers. US commanders will maintain the capability to report to higher US military authorities in addition to MNFCs. For matters perceived as illegal under US or international law, or outside the mandate of the mission to which the President has agreed, US commanders will first attempt resolution with the appropriate foreign MNFC. If issues remain unresolved, the US commanders refer the matters to higher US authorities.

1. Within the limits of OPCON, an MNFC cannot change the mission or deploy US forces outside the operational area agreed to by the President. Nor may the MNFC separate units, divide their supplies, administer discipline, promote anyone, or change the US force's internal organization.

2. Other MNF participants will normally exercise national or multinational command over their own forces on behalf of their governments.

(b) Tactical Control. TACON is another form of command authority exercised during multinational operations. It provides for the detailed (and usually local) direction and control of movements or maneuvers necessary to accomplish the missions or tasks assigned. The commander of the parent unit continues to exercise OPCON and ADCON over that unit unless otherwise specified in the establishing directive.

(c) Support. Supporting relationships may also be established among participating forces in multinational operations. US force commanders must be apprised of the opportunities, limitations, and/or conditions under which logistic support may be provided to forces of other nations.

(d) Coordinating Authority. In many cases, coordinating authority may be the only acceptable means of accomplishing a multinational mission. Coordinating authority is a consultation relationship between commanders, not an authority by which C2 may be exercised. Normally, it is more applicable to planning than to operations. Use of coordinating authority requires agreement among participants, as the commander exercising coordinating authority does not have the authority to resolve disputes. For this reason, its use during operations should be limited.

For further details concerning multinational operations, refer to JP 3-0, Joint Operations; JP 3-16, Multinational Operations; and the NATO series of allied joint doctrine publications.

Intentionally Blank

CHAPTER III
FUNCTIONS OF THE DEPARTMENT OF DEFENSE AND ITS MAJOR COMPONENTS

> *"It is a matter of record that the strategic direction of the war, as conducted by the Joint Chiefs of Staff, was fully as successful as were the operations which they directed...The proposals or the convictions of no one member were as sound, or as promising of success, as the united judgments and agreed decisions of all the members."*
>
> **Ernest J. King**
> **Fleet Admiral**
> **The US Navy at War, 1945**

SECTION A. DEPARTMENT OF DEFENSE

1. General

Unified action in carrying out the military component of NSS is accomplished through an organized defense framework. This chapter describes the components and their functions within that framework.

2. Organizations in the Department of Defense

a. Responsibility. SecDef is the principal assistant to the President in all matters relating to DOD. All functions in DOD and its component agencies are performed under the authority, direction, and control of SecDef.

b. DOD is composed of OSD, the Military Departments, the Joint Chiefs of Staff (JCS), the Joint Staff, the CCMDs, the Inspector General, agencies/bureaus, field activities, and such other offices, and commands established or designated by law, by the President, or by SecDef. The functions of the heads of these offices shall be as assigned by SecDef according to existing law.

3. Functions of the Department of Defense

As prescribed by higher authority, DOD will maintain and employ Armed Forces to:

a. Support and defend the Constitution of the US against all enemies, foreign and domestic.

b. Ensure, by timely and effective military action, the security of the US, its territories, and areas vital to its interest.

c. Uphold and advance the national policies and interests of the US.

Chapter III

4. Functions and Responsibilities Within the Department of Defense

a. The functions and responsibilities assigned to the Secretaries of the Military Departments, the Services, the JCS, the Joint Staff, and the CCMDs are carried out in such a manner as to achieve the following:

(1) Provide the best military advice to the President and SecDef.

(2) Effective strategic direction of the Armed Forces.

(3) Employment of the Armed Forces as a joint force.

(4) Integration of the Armed Forces into an effective and efficient team.

(5) Prevention of unnecessary duplication or overlapping capabilities among the Services by using personnel, intelligence, facilities, equipment, supplies, and services of all Services such that military effectiveness and economy of resources will thereby be increased.

(6) Coordination of Armed Forces operations to promote efficiency and economy and to prevent gaps in responsibility.

(7) Effective multinational operations and interagency, IGO, and NGO coordination.

b. **Development of Major Force Requirements.** Major force requirements to fulfill any specific primary function of an individual Service must be justified on the basis of existing or predicted need as recommended by the CJCS, in coordination with the JCS and CCDRs, and as approved by SecDef.

c. **Exceptions to Primary Responsibilities.** The development of special weapons and equipment and the provision of training equipment required by each of the Services are the responsibilities of the individual Service concerned unless otherwise directed by SecDef.

d. **Responsibility of a Service Chief on Disagreements Related to That Service's Primary Functions.** Each Service Chief is responsible for presenting to the CJCS any disagreement within the field of that Service's primary functions that has not been resolved. Any Service Chief may present unilaterally any issue of disagreement with another Service.

5. Executive Agents

a. SecDef or Deputy Secretary of Defense may designate a DOD executive agent (EA) and assign associated responsibilities, functions, and authorities within DOD. The head of a DOD component may be designated as a DOD EA. The DOD EA may delegate to a subordinate designee within that official's component the authority to act on that official's behalf for those DOD EA responsibilities, functions, and authorities assigned by SecDef

III-2

JP 1

or Deputy Secretary of Defense. Designation as EA confers no authority. The exact nature and scope of the DOD EA responsibilities, functions, and authorities shall be prescribed in the EA appointing document at the time of assignment and remain in effect until SecDef or Deputy Secretary of Defense revokes or supersedes them.

b. Responsibilities of an EA are established in DODD 5101.1, *Department of Defense Executive Agent,* and specific DODDs on specific EAs. Responsibilities of an EA include the following:

(1) Implement and comply with the relevant policies and directives of SecDef.

(2) Ensure proper coordination among Military Departments, the CCMDs, the JCS, the Joint Staff, the OSD, and DOD agencies and field activities.

(3) Issue directives to other DOD components and take action on behalf of SecDef.

(4) Make recommendations to SecDef for actions regarding the activity for which the EA was designated, including the manner and timing for dissolution of these responsibilities and duties.

(5) Perform such other duties and observe such limitations as set forth in the directive.

SECTION B. JOINT CHIEFS OF STAFF

6. Composition and Functions

a. The JCS consists of the CJCS; the Vice Chairman of the Joint Chiefs of Staff (VCJCS); the Chief of Staff, US Army; the Chief of Naval Operations; the Chief of Staff, US Air Force; the Commandant of the Marine Corps, and the CNGB. The Joint Staff supports the JCS and constitutes the immediate military staff of SecDef.

b. The CJCS is the principal military advisor to the President, NSC, HSC, and SecDef.

c. The other members of the JCS are military advisors to the President, NSC, HSC, and SecDef as specified below.

(1) A member of the JCS may submit to the CJCS advice or an opinion in disagreement with, or in addition to, the advice or opinion presented by the CJCS. If a member submits such advice or opinion, the CJCS shall present that advice or opinion to the President, NSC, or SecDef at the same time that he presents his own advice. The CJCS shall also, as he considers appropriate, inform the President, the NSC, or SecDef of the range of military advice and opinion with respect to any matter.

(2) The members of the JCS, individually or collectively, in their capacity as military advisors, shall provide advice on a particular matter when the President, NSC, HSC, or SecDef request such advice.

Chapter III

d. To the extent it does not impair independence in the performance of duties as a member of the JCS, Military Department Secretaries will inform their respective Service Chiefs regarding military advice rendered by members of the JCS on matters affecting their Military Departments.

e. The duties of the Service Chiefs and the CNGB as members of the JCS take precedence over all their other duties.

f. After first informing SecDef, a member of the JCS may make such recommendations to Congress relating to DOD as the member may consider appropriate.

g. When there is a vacancy, absence, or disability in the office of the CJCS, the VCJCS acts as and performs the duties of the CJCS until a successor is appointed or the absence or disability ceases.

h. When there is a vacancy in the offices of both the CJCS and VCJCS, or when there is a vacancy in one such office and in the absence or disability of the officer holding the other, the President will designate another member of the JCS to act as and perform the duties of the CJCS until a successor to the CJCS or VCJCS is appointed or the absence or disability of the CJCS or VCJCS ceases.

i. The Commandant of the Coast Guard may be invited by the CJCS or the Service Chiefs to participate in meetings or to discuss matters of mutual interest to the Coast Guard and the other Services.

7. Chairman of the Joint Chiefs of Staff

a. The CJCS is appointed by the President, with the advice and consent of the Senate, from the officers of the regular component of the United States Armed Forces.

b. The CJCS arranges for military advice, as appropriate, to be provided to all offices of SecDef.

c. While holding office, the CJCS outranks all other officers of the Armed Forces. The CJCS, however, may not exercise military command over the CCDRs, JCS, or any of the Armed Forces.

d. Subject to the authority, direction, and control of SecDef, the CJCS serves as the spokesman for the CCDRs, especially on the operational requirements of their commands. CCDRs will send their reports to the CJCS, who will review and forward the reports as appropriate to SecDef, subject to the direction of SecDef, so that the CJCS may better incorporate the views of CCDRs in advice to the President, the NSC, and SecDef. The CJCS also communicates the CCDRs' requirements to other elements of DOD.

e. The CJCS assists the President and SecDef in providing for the strategic direction of the Armed Forces. The CJCS transmits orders to the CCDRs as directed by the President or SecDef and coordinates all communications in matters of joint interest addressed to the CCDRs.

Functions of the Department of Defense and Its Major Components

(1) Develop and prepare strategic and deliberate plans.

(2) Advise on requirements, programs, and budget.

(3) Develop and establish doctrine, training, and education.

(4) Other matters.

(a) Provide for representation of the US on the Military Staff Committee of the UN in accordance with the USG law and policy.

(b) Perform such other duties as may be prescribed by law or by the President or SecDef.

(c) Conduct assessments and submit reports and budgets of the nature and magnitude of the strategic and military risks associated with executing the missions called for under the current NMS.

(d) Participate, as directed, in the preparation of multinational plans for military action in conjunction with the armed forces of other nations.

(e) Manage, for SecDef, the National Military Command System (NMCS), to meet the needs of SecDef and the JCS and establish operational policies and procedures for all components of the NMCS and ensure their implementation.

(f) Provide overall supervision of those DOD agencies and DOD field activities assigned to the CJCS by SecDef. Advise SecDef on the extent to which the program recommendations and budget proposals of a DOD agency or DOD field activity (DCMA, DISA, DIA, DLA, Missile Defense Agency, NGA, NSA, DTRA, and any other DOD agency designated as a CSA by SecDef), for which the CJCS has been assigned overall supervision, conform with the requirements of the Military Departments and of the CCMDs.

For further guidance on the CJCS functions, refer to Title 10, USC, Section 153.

8. Vice Chairman of the Joint Chiefs of Staff

a. The VCJCS is appointed by the President, by and with the advice and consent of the Senate, from the officers of the regular components of the United States Armed Forces.

b. The VCJCS holds the grade of general or admiral and outranks all other officers of the Armed Forces except the CJCS. The VCJCS may not exercise military command over the JCS, the CCDRs, or any of the Armed Forces.

c. The VCJCS performs the duties prescribed as a member of the JCS and such other duties and functions as may be prescribed by the CJCS with the approval of SecDef.

Chapter III

d. When there is a vacancy in the office of the CJCS, or in the absence or disability of the CJCS, the VCJCS acts as and performs the duties of the CJCS until a successor is appointed or the absence or disability ceases.

e. The VCJCS is a member of the Joint Nuclear Weapons Council, is the Vice Chairman of the Defense Acquisition Board, and may be designated by the CJCS to act as the Chairman of the Joint Requirements Oversight Council (JROC).

9. Joint Staff

a. The Joint Staff is under the exclusive authority, direction, and control of the CJCS. The Joint Staff will perform duties using procedures that the CJCS prescribes to assist the CJCS and the other members of the JCS in carrying out their responsibilities.

b. The Joint Staff includes officers selected in proportional numbers from the Army, Marine Corps, Navy, and Air Force. Coast Guard officers may also serve on the Joint Staff.

c. Selection of officers to serve on the Joint Staff is made by the CJCS from a list of officers submitted by the Services. Each officer whose name is submitted must be among those officers considered to be the most outstanding officers of that Service. The CJCS may specify the number of officers to be included on such a list.

d. After coordination with the other members of the JCS and with the approval of SecDef, the CJCS may select a Director, Joint Staff.

e. The CJCS manages the Joint Staff and its Director.

f. Per Title 10, USC, Section 155, the Joint Staff will not operate or be organized as an overall Armed Forces general staff and will have no executive authority. The Joint Staff is organized and operates along conventional staff lines to support the CJCS and the other members of the JCS in discharging their assigned responsibilities. In addition, the Joint Staff is the focal point for the CJCS to ensure that comments and concerns of the CCDRs and CSAs are well represented and advocated during all levels of coordination.

SECTION C. MILITARY DEPARTMENTS AND SERVICES

10. Common Functions of the Services and the United States Special Operations Command

a. Subject to the authority, direction, and control of SecDef and subject to the provisions of Title 10, USC, the Army, Marine Corps, Navy, and Air Force, under their respective Secretaries, are responsible for the functions prescribed in detail in DODD 5100.01, *Functions of the Department of Defense and Its Major Components*. Specific Service functions also are delineated in that directive.

b. USSOCOM is unique among the CCMDs in that it performs certain Service-like functions (in areas unique to SO) (Title 10, USC, Sections 161 and 167), including the following:

(1) Organize, train, equip, and provide combat-ready SOF to the other CCMDs and, when directed by the President or SecDef, conduct selected SO, usually in coordination with the GCC in whose AOR the SO will be conducted. USSOCOM's role in equipping and supplying SOF is generally limited to SO-peculiar equipment, materiel, supplies, and services.

(2) Develop strategy, doctrine, and tactics, techniques, and procedures for the conduct of SO, to include military information support operations (MISO) and CA forces. (Note: Joint doctrine is developed under the procedures approved by the CJCS.)

(3) Prepare and submit to SecDef program recommendations and budget proposals for SOF and other forces assigned to USSOCOM.

For additional information on SO, refer to JP 3-05, Special Operations.

SECTION D. COMBATANT COMMANDERS

11. General

a. GCCs are assigned a geographic AOR by the President with the advice of SecDef as specified in the UCP. Geographic AORs provide a basis for coordination by CCDRs. GCCs are responsible for the missions in their AOR, unless otherwise directed.

b. FCCs have transregional responsibilities and are normally supporting CCDRs to the GCC's activities in their AOR. FCCs may conduct operations as directed by the President or SecDef, in coordination with the GCC in whose AOR the operation will be conducted. The FCC may be designated by SecDef as the supported CCDR for an operation. The implementing order directing an FCC to conduct operations within a GCC's AOR will specify the CCDR responsible for mission planning and execution and appropriate command relationships.

c. Unless otherwise directed by the President or SecDef, the authority, direction, and control of the commander of a CCMD, with respect to the commands and the forces assigned to that command, are specified by law in Title 10, USC, Section 164. If a CCDR at any time considers the authority, direction, or control with respect to any of the commands or forces assigned to the CCDRs command to be insufficient to command effectively, the CCDR will promptly inform SecDef through the CJCS.

d. **Global Synchronizer.** SecDef or Deputy Secretary of Defense may assign a CCDR global synchronizer responsibilities. A global synchronizer is the CCDR responsible for the alignment of specified planning and related activities of other CCMDs, Services, DOD agencies and activities, and as directed, appropriate USG departments and agencies within an established, common framework to facilitate coordinated and decentralized execution across geographic and other boundaries. The global

Chapter III

synchronizer's role is to align and harmonize plans and recommend sequencing of actions to achieve the strategic end states and objectives of a GCP.

e. The authority that CCDRs and Service Secretaries (when operating as force providers as designated by SecDef) may exercise over assigned RC forces when not on active duty or when on active duty for training is known as training and readiness oversight (TRO). See Department of Defense Instruction (DODI) 1215.06, *Uniform Reserve, Training, and Retirement Categories,* and Chapter V, "Joint Command and Control," Paragraph 10, "Command of National Guard and Reserve Forces."

12. Geographic Combatant Command Responsibilities

a. Based on the President's UCP, the Commanders, US Central Command, USEUCOM, United States Pacific Command (USPACOM), US Southern Command, US Africa Command, and USNORTHCOM, are each assigned a geographic AOR within which their missions are accomplished with assigned and/or attached forces. Forces under the direction of the President or SecDef may conduct operations from or within any geographic area as required for accomplishing assigned tasks, as mutually agreed by the CCDRs concerned or as specifically directed by the President or SecDef. Some responsibilities of these GCCs are to:

(1) Detect, deter, and prevent attacks against the US, its territories and bases, and employ appropriate force should deterrence fail.

(2) Carry out assigned missions and tasks, and plan for and execute military operations, as directed.

(3) Assign tasks to and direct coordination among subordinate commands.

(4) Maintain the security of and carry out force protection and personnel recovery responsibilities for the command, including assigned or attached commands, forces, and assets.

(5) Plan, conduct, and assess security cooperation activities.

(6) Plan and, as appropriate, conduct evacuation and protection of US citizens and nationals and designated other persons.

(7) Provide US military representation to international and US national agencies unless otherwise directed.

(8) Provide the single point of contact on military matters within the AOR.

b. The Commander, United States Northern Command (CDRUSNORTHCOM), is responsible for:

(1) Providing support to civil authorities, as directed.

Functions of the Department of Defense and Its Major Components

(2) Providing chemical, biological, radiological, and nuclear consequence management (CBRN CM) assistance and support to US and allied partner authorities, as directed, within US territories and protectorates and the USNORTHCOM AOR.

(3) CDRUSNORTHCOM is also designated the Commander, US Element, NORAD, and will be designated Commander or Deputy Commander of NORAD. A binational command of the US and Canada, NORAD is responsible for aerospace warning and control and maritime warning for Canada, Alaska, Puerto Rico, the US Virgin Islands, the continental US, the air defense identification zone, and the air and maritime approaches. Through NORAD, the commander answers to both the US and Canadian governments.

(4) Planning, organizing, and as directed, executing HD operations within the USNORTHCOM AOR in concert with missions performed by Commander, NORAD.

(5) Synchronizing planning for DOD efforts in support of the USG response to pandemic influenza.

c. The Commander, USPACOM, is responsible for:

(1) Providing support to civil authorities, as directed.

(2) Providing CBRN CM assistance and support to US and allied partner authorities, as directed, within US territories and protectorates and the USPACOM AOR.

(3) Planning, organizing, and as directed, executing HD operations within the USPACOM AOR.

(4) Synchronizing planning for DOD efforts in support of the USG response to pandemic influenza and infectious disease.

13. **Functional Combatant Command Responsibilities**

a. CDRUSSOCOM is an FCC who exercises COCOM of all assigned AC and mobilized RC SOF minus US Army Reserve CA and MISO forces. When directed, CDRUSSOCOM provides US-based SOF to a GCC who exercises COCOM of assigned and OPCON of attached SOF through a commander of a theater SO command or a joint SO task force in a specific operational area or to prosecute SO in support of a theater campaign or other operations. SOF are those forces identified in Title 10, USC, Section 167, or those units or forces that have since been designated as SOF by SecDef, and they are those AC and RC forces of the Services specifically organized, trained, and equipped to conduct and support SO. When directed, CDRUSSOCOM can establish and employ a JTF as the supported commander. In addition to functions specified in Title 10, USC, Section 167, CDRUSSOCOM is responsible to:

(1) Serve as the SOF joint force provider.

III-9

Chapter III

(2) Integrate and coordinate DOD MISO capabilities to enhance interoperability and support US Strategic Command's IO responsibilities and other CCDRs' MISO planning and execution.

(3) Synchronize planning for global operations against terrorist networks in coordination with other CCDRs, the Services, and as directed, appropriate USG departments and agencies.

(4) Train SO force, including developing recommendations to the CJCS regarding strategy, doctrine, tactics, techniques, and procedures for the joint employment of SOF.

b. The Commander, US Strategic Command, is an FCC who is responsible to:

(1) Maintain primary responsibility among CCDRs to support the national objective of strategic deterrence.

(2) Provide integrated global strike planning and C2 support of theater and national objectives and exercising C2 of selected missions as directed.

(3) Synchronize planning for global missile defense in coordination with other CCDRs, the Services, and as directed, appropriate USG departments and agencies.

(4) Plan, integrate, and coordinate ISR in support of strategic and global operations.

(5) Provide planning, training, and contingent electronic warfare support.

(6) Synchronize planning for DOD combating WMD efforts in coordination with other CCDRs, the Services, and as directed, appropriate USG departments and agencies.

(7) Plan and conduct space operations.

(8) Synchronize planning for cyberspace operations.

(9) Provide in-depth analysis and precision targeting for selected networks and nodes.

c. The Commander, US Transportation Command, is an FCC who is responsible to:

(1) Provide common-user and commercial air, land, and maritime transportation, terminal management, and aerial refueling to support global deployment, employment, sustainment, and redeployment of US forces.

(2) Serve as the mobility joint force provider.

(3) Provide DOD global patient movement, in coordination with GCCs, through the Defense Transportation Network.

III-10   JP 1

(4) Serve as the Distribution Process Owner. Synchronize planning for global distribution operations in coordination with other CCMDs, the Services, and as directed, appropriate government departments and agencies.

For further detail concerning CCDRs' assigned responsibilities, refer to the UCP.

14. Statutory Command Authority

Title 10, USC, Section 164, outlines seven command functions of CCDRs. The functions are shown in Figure III-1.

15. Authority Over Subordinate Commanders

Unless otherwise directed by the President or SecDef, commanders of the CCMDs exercise authority over subordinate commanders as follows:

a. Commanders of commands and forces assigned to a CCDR are under the authority, direction, and control of, and are responsible to, the CCDR on all matters for which the CCDR has been assigned authority under Section 164, subsection (c) of Title 10, USC.

b. The commander of a command or force assigned to a CCDR will communicate with other elements of DOD on any matter for which the commander of the CCMD has been assigned authority in accordance with procedures, if any, established by the CCDR.

c. Other elements of DOD will communicate with the commander of a command or force referred to in Paragraph a. above on any matter for which the CCDR has been

---

Command Functions of a Combatant Commander

- Giving authoritative direction to subordinate commands and forces necessary to carry out missions assigned to the command, including authoritative direction over all aspects of military operations, joint training, and logistics.

- Prescribing the chain of command to the commands and forces within the command.

- Organizing commands and forces within that command as necessary to carry out missions assigned to the command.

- Employing forces within that command as necessary to carry out missions assigned to the command.

- Assigning command functions to subordinate commanders.

- Coordinating and approving those aspects of administration, support (including control of resources and equipment, internal organization, and training), and discipline necessary to carry out missions assigned to the command.

- Exercising the authority with respect to selecting subordinate commanders, selecting combatant command staff, suspending subordinates, and convening courts-martial as delineated in Title 10, US Code, Section 164.

Figure III-1. Command Functions of a Combatant Commander

Chapter III

assigned authority under Section 164, subsection (c) of Title 10, USC, in accordance with procedures, if any, established by the CCDR.

d. If directed by the CCDR, the commander of a command or force referred to in Paragraph a. above shall advise the CCDR of all communications to and from other elements of DOD on any matter for which the commander of the CCMD has not been assigned authority under Section 164, subsection (c) of Title 10, USC.

e. Commanders of commands and forces assigned to a CCDR with bilateral or multilateral planning requirements coordinate operation planning efforts with that CCDR, in order to synchronize that plan with the CCDR's campaign plans and regional strategy.

16. Department of Defense Agencies

DOD agencies are organizational entities of DOD established by SecDef under Title 10, USC, to perform a supply or service activity common to more than one Military Department.

For more information on DOD agencies, see DODD 3000.06, Combat Support Agencies.

III-12   JP 1

CHAPTER IV
JOINT COMMAND ORGANIZATIONS

"...success rests in the vision, the leadership, the skill and the judgment of the professionals making up command and staff groups..."

Dwight D. Eisenhower
Crusade in Europe, 1948

SECTION A. ESTABLISHING UNIFIED AND SUBORDINATE JOINT COMMANDS

1. General

Joint forces are established at three levels: unified CCMDs, subordinate unified commands, and JTFs.

a. **Authority to Establish.** In accordance with the National Security Act of 1947 and Title 10, USC, and as described in the UCP, CCMDs are established by the President, through SecDef, with the advice and assistance of the CJCS. Commanders of unified CCMDs may establish subordinate unified commands when so authorized by SecDef through the CJCS. JTFs can be established by SecDef, a CCDR, subordinate unified commander, or an existing JTF commander.

b. **Basis for Establishing Joint Forces.** Joint forces can be established on either a geographic area or functional basis.

(1) Geographic Area. Establishing a joint force on a geographic area basis is the most common method to assign responsibility for continuing operations. The title of the areas and their delineation are prescribed in the establishing directive. Note: Only GCCs are assigned AORs. GCCs normally assign subordinate commanders an operational area from within their assigned AOR.

For further information on operational areas, refer to JP 3-0, Joint Operations.

(a) The UCP contains descriptions of the geographic boundaries assigned to GCCs. These geographic AORs do not restrict accomplishment of assigned missions; CCDRs may operate forces wherever required to accomplish their missions. The UCP provides that, unless otherwise directed by SecDef, when significant operations overlap the boundaries of two GCCs' AORs, a JTF will be formed. Command of this JTF will be determined by SecDef and forces transferred to the JTF commander through a CCDR, including delegation of appropriate command authority over those forces.

(b) Each GCC and subordinate JFC will be kept apprised of the presence, mission, movement, and duration of stay of transient forces within the operational area. The subordinate JFC also will be apprised of the command channels under which these transient forces will function. The authority directing movement or permanent location of transient forces is responsible for providing this information.

IV-1

Chapter IV

(c) Forces not assigned or attached to a GCC or attached to a subordinate JFC often are assigned missions that require them to cross boundaries. In such cases, it is the duty of the JFC to assist the operations of these transient forces to the extent of existing capabilities and consistent with other assigned missions. The JFC may be assigned specific responsibilities with respect to installations or activities exempted from their control, such as logistic support or area defense, particularly if adversary forces should traverse the operational area to attack the exempted installation or activity. GCC force protection policies take precedence over all force protection policies or programs of any other DOD component deployed in that GCC's AOR and not under the security responsibility of DOS. The GCC or a designated representative (e.g., a JTF or component commander) shall delineate the force protection measures for all DOD personnel not under the responsibility of DOS.

(d) Transient forces within the assigned AOR of a CCDR are subject to that CCDR's orders in some instances (e.g., for coordination of emergency defense, force protection, or allocation of local facilities).

(2) Functional. Sometimes a joint force based solely on military functions without respect to a specific geographic region is more suitable to fix responsibility for certain types of continuing operations (e.g., the unified CCMDs for transportation, SO, and strategic operations). The commander of a joint force established on a functional basis is assigned a functional responsibility by the establishing authority.

(a) When defining functional responsibilities, the focus should be on the mission and objective(s) or service provided. The title of the functional responsibility and its delineation are prescribed in the establishing directive.

(b) The missions or tasks assigned to the commander of a functional command may require that certain installations and activities of that commander be exempt, partially or wholly, from the command authority of a GCC in whose area they are located or within which they operate. Such exemptions must be specified by the authority that establishes the functional command. Such exemptions do not relieve the commanders of functional commands of the responsibility to coordinate with the affected GCC.

c. **Organizing Joint Forces.** A JFC has the authority to organize assigned or attached forces with specification of OPCON to best accomplish the assigned mission based on his intent, the CONOPS, and consideration of Service organizations (see Figure IV-1). The organization should be sufficiently flexible to meet the planned phases of the contemplated operations and any development that may necessitate a change in plan. The JFC will establish subordinate commands, assign responsibilities, establish or delegate appropriate command relationships, and establish coordinating instructions for the component commanders. Sound organization should provide for unity of command, centralized planning and direction, and decentralized execution. Unity of effort is necessary for effectiveness and efficiency. Centralized planning and direction is essential for controlling and coordinating the efforts of the forces. Decentralized execution is essential because no one commander can control the detailed actions of a large number of units or individuals. When organizing joint forces with MNFs, simplicity and clarity are critical. Complex or

IV-2   JP 1

Joint Command Organizations

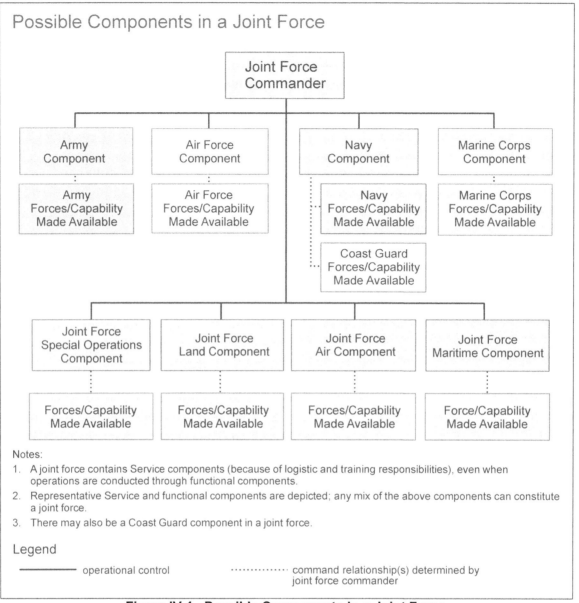

Figure IV-1. Possible Components in a Joint Force

unclear command relationships or organization are counterproductive to developing synergy among MNFs.

(1) The composition of the JFC's staff will reflect the composition of the joint force to ensure that those responsible for employing joint forces have a thorough knowledge of the capabilities and limitations of assigned or attached forces.

(2) All joint forces include Service components, because administrative and logistic support for joint forces are provided through Service components. Service forces may be assigned or attached to subordinate joint forces without the formal creation of a respective Service component command of that joint force. In the absence of a formally designated Service component commander, the senior Service commander assigned or

Chapter IV

attached to the subordinate JFC will accomplish Service ADCON. This relationship is appropriate when stability, continuity, economy, ease of long-range planning, and the scope of operations dictate organizational integrity of Service forces for conducting operations.

See Paragraph 8, "Service Component Commands," for more information on Service component commands.

(3) The JFC can establish functional component commands to conduct operations. Functional component commands are appropriate when forces from two or more Military Departments must operate within the same mission area or physical domain or there is a need to accomplish a distinct aspect of the assigned mission. Joint force land, air, maritime, and SO component commanders are examples of functional components. (Note: Functional component commanders are component commanders of a joint force and do not constitute a "joint force command" with the authorities and responsibilities of a JFC, even when employing forces from two or more Military Departments.) When a functional component command employs forces from more than one Service, the functional component commander's staff should include Service representatives from each of the employed Service forces to aid in understanding those Service capabilities and maximizing the effective employment of Service forces. Joint staff billets for needed expertise and individuals to fill those billets should be identified. Those individuals should be used when the functional component command is formed for exercises, contingency planning, or actual operations.

See Paragraph 9, "Functional Component Commands," for more information on functional component commands.

(4) Normally, joint forces are organized with a combination of Service and functional component commands with operational responsibilities. Joint forces organized with Army, Marine Corps, Navy, and Air Force components may have SOF (if assigned) organized as a functional component. The JFC defines the authority, command relationships, and responsibilities of the Service and functional component commanders; however, the Service responsibilities (i.e., administrative and logistics) of the components must be given due consideration by the JFC.

(5) The JFC has full authority to assign missions, redirect efforts, and direct coordination among subordinate commanders. JFCs should allow Service tactical and operational assets and groupings to function generally as they were designed and organized (e.g., carrier strike group; Marine air-ground task force [MAGTF]; US Air Force air and space expeditionary task forces; US Army corps, divisions, brigade combat teams; and SO JTF). The intent is to meet the needs of the JFC while maintaining the tactical and operational integrity of the Service organizations. The following policy for C2 of United States Marine Corps tactical air (TACAIR) recognizes this and deals with MAGTF aviation during sustained operations ashore.

(a) The MAGTF commander will retain OPCON of organic air assets. The primary mission of the MAGTF aviation combat element is the support of the MAGTF

IV-4   JP 1

Joint Command Organizations

ground combat element. During joint operations, the MAGTF air assets normally will be in support of the MAGTF mission. The MAGTF commander will make sorties available to the JFC, for tasking through the joint force air component commander (JFACC), for air defense, long-range interdiction, and long-range reconnaissance. Sorties in excess of MAGTF direct support requirements will be provided to the JFC for tasking through the JFACC for the support of other components of the joint force or the joint force as a whole.

(b) Nothing herein shall infringe on the authority of the GCC or subordinate JFC in the exercise of OPCON to assign missions, redirect efforts (e.g., the reapportionment and/or reallocation of any MAGTF TACAIR sorties when it has been determined by the JFC that they are required for higher priority missions), and direct coordination among the subordinate commanders to ensure unity of effort in accomplishment of the overall mission, or to maintain integrity of the force.

> **Note: Sorties provided for air defense, long-range interdiction, and long-range reconnaissance are not "excess" sorties and will be covered in the air tasking order. These sorties provide a distinct contribution to the overall joint force effort. The joint force commander must exercise integrated control of air defense, long-range reconnaissance, and interdiction aspects of the joint operation or theater campaign. Excess sorties are in addition to these sorties.**

2. Unified Combatant Command

a. **Criteria for Establishing a Unified Combatant Command.** A unified CCMD is a **command with broad continuing missions under a single** commander and composed of significant assigned components of two or more Military Departments that is established and so designated by the President through SecDef and with the advice and assistance of the CJCS. When either or both of the following criteria apply generally to a situation, a unified CCMD normally is required to ensure unity of effort.

(1) A broad continuing mission exists, requiring execution by significant forces of two or more Military Departments and necessitating a single strategic direction.

(2) Any combination of the following exists and significant forces of two or more Military Departments are involved:

(a) A large-scale operation requiring positive control of tactical execution by a large and complex force;

(b) A large geographic or functional area requiring single responsibility for effective coordination of the operations therein; and/or

(c) Necessity for common use of limited logistic means.

b. The commander of a unified CCMD normally will adapt the command structure to exercise command authority through the commander of a **subunified command, JTF, Service component, or functional component.** Alternatively, the commander of a unified

Chapter IV

CCMD may choose to exercise command authority directly through the commander of a **single-Service force** (e.g., task force, task group, MAGTF for a noncombatant evacuation operation) or a **specific operational force** (e.g., SOF for an SO core operation), who, because of the mission assigned and the urgency of the situation, must remain immediately responsive to the CCDR. The commander of a unified CCMD normally assigns missions requiring a single-Service force to a Service component commander. These six options (shown in Figure IV-2) do not in any way limit the commander's authority to organize subordinate commands and exercise command authority over assigned forces as they see fit.

c. The **commander of a unified CCMD should not act concurrently as the commander of a subordinate command.** For example, the commander of a unified CCMD should not act as a functional component commander without prior approval of SecDef.

d. **Primary Responsibilities of the Commander of a Unified Combatant Command.** CCDRs are responsible for the development and production of joint plans and orders. During peacetime, they act to deter war through military engagement and security cooperation activities and prepare to execute other missions that may be required. During a conflict/combat, they plan and conduct campaigns and major operations to accomplish assigned missions. Unified CCMD responsibilities include the following:

(1) Planning and conducting military operations in response to crises, to include the security of the command and protection of the US and its territories and bases against attack or hostile incursion. The JSCP tasks the CCDRs to prepare joint contingency plans that may be one of four increasing levels of detail: commander's estimate, basic plan, concept plan, or operation plan.

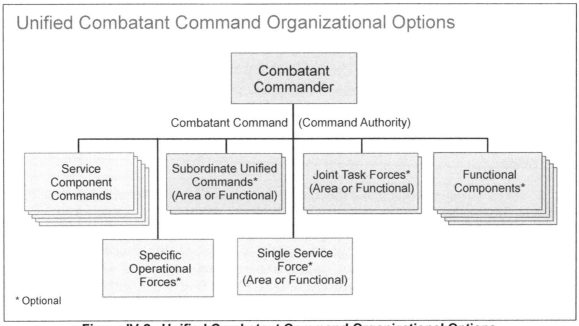

Figure IV-2. Unified Combatant Command Organizational Options

Joint Command Organizations

For further detail concerning joint planning, refer to JP 5-0, Joint Operation Planning, *and* Chairman of the Joint Chiefs of Staff Manual (CJCSM) 3122.01A, Joint Operation Planning and Execution System (JOPES), Volume I (Planning Policies and Procedures).

(2) Maintaining the preparedness of the command to carry out missions assigned to the command.

(3) Carrying out assigned missions, tasks, and responsibilities.

(4) Assigning tasks to, and directing coordination among, the supporting CCDRs and the subordinate commands to ensure unity of effort in the accomplishment of the assigned missions.

(5) Communicating directly with the following:

(a) The Service Chiefs on single-Service matters as appropriate.

(b) The CJCS on other matters, including the preparation of strategic, joint operation, and logistic plans; strategic and operational direction of assigned forces; conduct of combat operations; and any other necessary function of command required to accomplish the mission.

(c) SecDef, in accordance with applicable directives.

(d) Subordinate elements, including the development organizations of the DOD agency or the Military Department directly supporting the development and acquisition of the CCDR's C2 system in coordination with the director of the DOD agency or Secretary of the Military Department concerned.

(6) Keeping the CJCS promptly advised of significant events and incidents that occur in the functional area or area of operations, particularly those incidents that could create national or international repercussions.

(7) Establish relationships with the CNGB to advise on National Guard matters pertaining to their CCMD missions and support planning and coordination for such activities as requested by the CJCS.

e. **Authority of the Commander of a Unified Combatant Command in an Emergency**

(1) In the event of a major emergency in the GCC's AOR requiring the use of all available forces, the GCC (except for CDRUSNORTHCOM) may temporarily assume OPCON of all forces in the assigned AOR, including those of another command, but excluding those forces scheduled for or actually engaged in the execution of specific operational missions under joint plans approved by SecDef that would be interfered with by the contemplated use of such forces. CDRUSNORTHCOM's authority to assume OPCON during an emergency is limited to the portion of USNORTHCOM's AOR outside the US. CDRUSNORTHCOM must obtain SecDef approval before assuming OPCON of

Chapter IV

forces not assigned to USNORTHCOM within the US. The commander determines when such an emergency exists and, on assuming OPCON over forces of another command, immediately advises the following individual(s) of the nature and estimated duration of employment of such forces.

(a) The CJCS.

(b) The appropriate operational commanders.

(c) The Service Chief of the forces concerned.

(2) The authority to assume OPCON of forces in the event of a major emergency will not be delegated.

(3) Unusual circumstances in wartime, emergencies, or crises (such as a terrorist incident) may require a GCC to directly exercise COCOM through a shortened chain of command to forces assigned for the purpose of resolving the crisis. Additionally, the GCC can assume COCOM, in the event of war or an emergency that prevents control through normal channels, of security assistance organizations within the GCC's general geographic AOR, or as directed by SecDef. All commanders bypassed in such exceptional command arrangements will be advised of directives issued to and reports sent from elements under such arrangements. Such arrangements will be terminated as soon as practicable, consistent with mission accomplishment.

f. **GCC Authority for Force Protection Outside the US**

(1) GCCs shall exercise authority for force protection over all DOD personnel (including their dependents) assigned, attached, transiting through, or training in the GCC's AOR, except for those for whom the chief of mission retains security responsibility.

(2) Transient forces do not come under the authority of the GCC solely by their movement across operational area boundaries, except when the GCC is exercising TACON authority for force protection purposes or in the event of a major emergency as stated in Paragraph 2.e.(1).

(3) This force protection authority enables GCCs to change, modify, prescribe, and enforce force protection measures for covered forces.

For further detail concerning the force protection authority of the GCCs, refer to DODD 2000.12, DOD Antiterrorism (AT) Program, and JP 3-07.2, Antiterrorism.

g. **GCC Authority for Exercise Purposes.** Unless otherwise specified by SecDef, and with the exception of the USNORTHCOM AOR, a GCC has TACON for exercise purposes whenever forces not assigned to that CCDR undertake exercises in that GCC's AOR. TACON begins when the forces enter the AOR. In this context, TACON provides directive authority over exercising forces for purposes relating to force protection and to that exercise only; it does not authorize operational employment of those forces.

h. **Assumption of Interim Command.** In the temporary absence of a CCDR from the command, interim command will pass to the deputy commander. If a deputy commander has not been designated, interim command will pass to the next senior officer present for duty who is eligible to exercise command, regardless of Service affiliation.

3. **Specified Combatant Command**

There are currently no specified CCMDs designated. Because the option for the President through SecDef to create a specified CCMD still exists in Title 10, USC, Section 161, the following information is provided. A specified CCMD is a command that has broad continuing missions and is established by the President, through SecDef, with the advice and assistance of the CJCS (see Figure IV-3).

a. **Composition.** Although a specified CCMD normally is composed of forces from one Military Department, it may include units and staff representation from other Military Departments.

b. **Transfer of Forces from Other Military Departments.** When units of other Military Departments are transferred (assigned or attached) to the commander of a specified CCMD, the purpose and duration of the transfer normally will be indicated. Such transfer does not constitute the specified CCMD as a unified CCMD or a JTF. If the transfer is major and of long duration, a unified CCMD normally would be established in lieu of a specified CCMD.

c. **Authority and Responsibilities.** The commander of a specified CCMD has the same authority and responsibilities as the commander of a unified CCMD, except that no authority exists to establish subordinate unified commands.

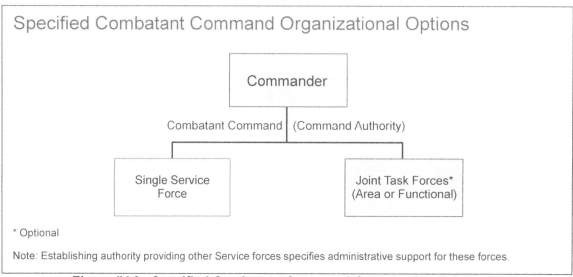

Figure IV-3. Specified Combatant Command Organizational Options

Chapter IV

4. Subordinate Unified Command

When authorized by SecDef through the CJCS, commanders of unified CCMDs may establish subordinate unified commands (also called subunified commands) to conduct operations on a continuing basis in accordance with the criteria set forth for unified CCMDs (see Figure IV-4). A subordinate unified command (e.g., United States Forces Korea) may be established on a geographical area or functional basis. Commanders of subordinate unified commands have functions and responsibilities similar to those of the commanders of unified CCMDs and exercise OPCON of assigned commands and forces and normally over attached forces within the assigned joint operations area or functional area. The commanders of components or Service forces of subordinate unified commands have responsibilities and missions similar to those for component commanders within a unified CCMD. The Service component or Service force commanders of a subordinate unified command normally will communicate directly with the commanders of the Service component command of the unified CCMD on Service-specific matters and inform the commander of the subordinate unified command as that commander directs.

5. Joint Task Force

As shown in Figure IV-5, a JTF is a joint force that is constituted and so designated by SecDef, a CCDR, a subordinate unified commander, or an existing JTF commander.

a. A JTF may be established on a geographical area or functional basis when the mission has a specific limited objective and does not require overall centralized control of logistics. The mission assigned to a JTF requires execution of responsibilities involving a joint force on a significant scale and close integration of effort, or requires coordination

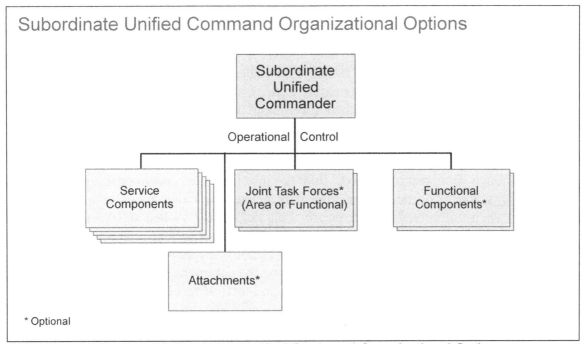

Figure IV-4. Subordinate Unified Command Organizational Options

Joint Command Organizations

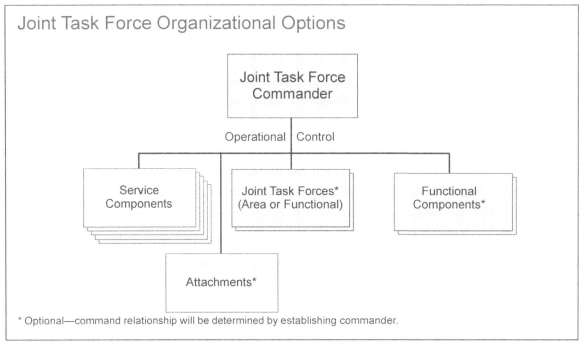

Figure IV-5. Joint Task Force Organizational Options

within a subordinate area or local defense of a subordinate area. The establishing authority dissolves a JTF when the purpose for which it was created has been achieved or when it is no longer required.

b. The authority establishing a JTF designates the commander, assigns the mission, designates forces, and delegates command authorities.

c. Based on the decision of the establishing JFC, the commander of a JTF exercises OPCON over assigned (and normally over attached) forces, or may exercise TACON over attached forces. The JTF commander establishes command relationships between subordinate commanders and is responsible to the establishing commander for the proper employment of assigned and attached forces and for accomplishing such operational missions as may be assigned. JTF commanders also are responsible to the establishing commander for the conduct of joint training of assigned forces.

d. Although not recommended, due to the need to concentrate on JTF-level considerations, the JTF commander also may act as the commander of a subordinate command, when authorized by the establishing authority. When this is the case, the commander also has the responsibilities associated with the subordinate command for the forces belonging to the parent Service. The JTF establishing authority should consider that dual-hatting a commander also means dual-hatting the commander's staff, which can result in forcing the staff to operate at the operational and tactical levels simultaneously.

e. The JTF commander will have a **joint staff** with appropriate members in key positions of responsibility from each Service or functional component having significant forces assigned to the command.

Chapter IV

For further detail concerning JTFs, refer to JP 3-33, Joint Task Force Headquarters.

SECTION B. COMMANDER, STAFF, AND COMPONENTS OF A JOINT FORCE

6. Commander Responsibilities

a. **Responsibilities of a JFC.** Although specific responsibilities will vary, a JFC possesses the following general responsibilities:

(1) Provide a clear commander's intent and timely communication of specified tasks, together with any required coordinating and reporting requirements. Tasks should be realistic yet leave subordinate commanders flexibility in their CONOPS and the ability to take the initiative as opportunities arise.

(2) Transfer forces and other capabilities to designated subordinate commanders for accomplishing assigned tasks.

(3) Provide all available information to subordinate JFCs and component commanders that affect their assigned missions and objectives.

(4) Delegate authority to subordinate JFCs and component commanders commensurate with their responsibilities.

b. **Responsibilities of a Subordinate Commander.** In addition to other responsibilities that change according to circumstances, all subordinate commanders possess the general responsibilities to provide for the following:

(1) The accomplishment of missions or tasks assigned by the plans and orders of the superior. Authority normally is given to the subordinate commander to select the methodology for accomplishing the mission; however, this may be limited by establishing directives issued by the superior JFC to ensure effective joint operations. Any departure from the plan by a subordinate commander should, if possible, be coordinated with other commanders prior to departure from the plan. In addition, the departure must be communicated as soon as practicable to the superior.

(2) Advice to the superior JFC regarding employment possibilities of and consequences to achieving the subordinate commander mission objectives, cooperation with appropriate government and nongovernmental agencies, and other matters of common concern.

(3) Timely information to the superior JFC relating to the subordinate commander's situation and progress.

c. **Responsibilities of Other Commanders.** Commanders who share a common higher commander or a common boundary are responsible for the following:

(1) Consider the impact of one's own actions or inactions on other friendly forces.

IV-12   JP 1

Joint Command Organizations

(2) Provide timely information to other JFCs regarding one's own intentions and actions, as well as those of nonmilitary agencies or of the adversary, which may influence other activity.

(3) Support other JFCs as required by the common aim and the situation.

(4) Coordinate the support provided and received.

d. **Responsibilities of Deputy Commanders.** Deputy JFCs in a joint force may be designated as the JFC's principal assistant available to replace the JFC, if needed. Normally, the deputy commander is not a member of the same Service as the JFC. The deputy JFC's responsibilities may include the following:

(1) Performing special duties (e.g., chair the joint targeting coordination board, cognizance of liaison personnel reporting to the joint force headquarters, interorganizational coordination).

(2) Working with the components to keep the JFC updated.

(3) Recommending refinements in the relationship with and between the components to provide the most efficient and effective command relationships.

(4) Tracking the JFC's critical information requirements to ensure compliance.

7. Staff of a Joint Force

a. **General.** A joint staff should be established for commands composed of more than one Military Department. The staff of the commander of a CCMD, subordinate unified command, or JTF must be composed of Service members that constitute significant elements of the joint force. Positions on the staff should be divided so that Service representation and influence generally reflect the Service composition of the force.

(1) A JFC is authorized to organize the staff and assign responsibilities to individual Service members assigned to the staff as deemed necessary to accomplish assigned missions.

(2) A joint staff should be reasonably balanced as to numbers, experience, influence of position, and rank of the Service members concerned. The composition of a joint staff should be commensurate with the composition of forces and the character of the contemplated operations to ensure that the staff understands the capabilities, needs, and limitations of each element of the force. The number of personnel on a joint staff should be kept to the minimum consistent with the task to be performed.

(3) Each person assigned to serve on a joint staff will be responsible to the JFC and should have thorough knowledge of the JFC's policies.

(4) Recommendations of any member of the staff receive appropriate consideration.

Chapter IV

(5) The degree of authority to act in the name of and for the JFC is a matter to be specifically prescribed by the JFC.

(6) Orders and directives from a higher to a subordinate command should be issued in the name of the JFC of the higher command to the JFC of the immediate subordinate command and not directly to elements of that subordinate command. Exceptions may sometimes be required under certain emergency or crisis situations. C2 of nuclear forces is an example of one such exception.

(7) To expedite the execution of orders and directives and to promote teamwork between commands, a JFC may authorize the command's staff officers to communicate directly with appropriate staff officers of other commands concerning the details of plans and directives that have been received or are to be issued.

(8) Each staff division must coordinate its actions and planning with the other staff directorates concerned and keep them informed of actions taken and the progress achieved. Each general or joint staff directorate is assigned responsibility for a particular type of problem and subject, and for coordinating the work of the special staff divisions and other staff elements that relate to that problem or subject.

(9) Joint staff directorates and special staff sections should be limited to those functions for which the JFC is responsible or that require the JFC's general supervision in the interest of unity of effort.

(10) The authority that establishes a joint force should make the provisions for furnishing necessary personnel for the JFC's staff.

b. **Staff Organization.** The staff organization should generally conform to the principles established in this section.

(1) Principal Staff Officer. The chief of staff (COS) functions as the principal staff officer, assistant, and advisor to the JFC. The COS coordinates and directs the work of the staff directorates. One or more deputies to the COS and a secretary of the staff may be provided to assist the COS in the performance of assigned duties. A deputy COS normally should be from a Service other than that of the COS. The secretary of the staff is the executive in the office of the COS and is responsible for routing and forwarding correspondence and papers and maintaining office records.

(2) Personal Staff Group of the Commander. The JFC's personal staff performs duties prescribed by the JFC and is responsible directly to the JFC. This group, composed of aides to the JFC and staff officers handling special matters over which the JFC exercises close personal control will include a staff judge advocate, political advisor, PA officer, inspector general, provost marshal, chaplain, surgeon, historian, and others as directed. JFCs should be aware that participation in multinational operations may affect the size and responsibilities of this group.

(3) Special Staff Group. The special staff group consists of representatives of technical or administrative services and can include representatives from government or

IV-14   JP 1

Joint Command Organizations

nongovernmental agencies. The general functions of the special staff include furnishing technical, administrative, and tactical advice and recommendations to the JFC and other staff officers; preparing the parts of plans, estimates, and orders in which they have primary interest; and coordinating and supervising the activities for which each staff division is responsible. Because the headquarters of a joint force is concerned primarily with broad operational matters rather than with technical problems associated with administration and support of assigned and/or attached forces, this group should be small to avoid unnecessary duplication of corresponding staff sections or divisions within the Service component headquarters. When a JFC's headquarters is organized without a special staff group, the officers who might otherwise compose the special staff group may be organized as branches of the divisions of the joint staff or as additional joint staff divisions.

(4) **Joint Force Staff Directorates.** The joint staff group typically is made up of staff directorates corresponding to the major functions of command, such as personnel, intelligence, operations, logistics, plans, and communications systems. The head of each staff directorate may be designated as a director or as an assistant COS. The directors or assistant COSs provide staff supervision for the JFC of all activities pertaining to their respective functions.

(5) **Liaison Officers and/or Agency Representatives.** Liaisons or representatives from various higher, lower, and adjacent organizations, DOD agencies, and non-DOD entities are normally spread throughout the joint force staff and not grouped as a separate entity. However, considering the increasing complexity of joint and/or interagency coordination, the JFC may decide to consolidate, at least administratively, liaisons and representatives in a single interagency office, and then provide them to specific directorates or components where they would best be employed and of value to their parent agency or command. The administration and assignment of liaison officers is normally under the cognizance of the deputy JFC or the COS.

8. Service Component Commands

a. A Service component command, assigned to a CCDR, consists of the Service component commander and the Service forces (such as individuals, units, detachments, and organizations, including the support forces) that have been assigned to that CCDR. Forces assigned to CCDRs are identified in the GFMIG signed by SecDef. Service components can only be assigned under COCOM to one CCDR. However, Service component commanders may support multiple CCDRs in a supporting relationship, while not assigned to any of the supported CCDRs.

b. **Designation of Service Component Commanders.** With the exception of the commander of a CCMD and members of the command's joint staff, the senior officer of each Service assigned to a CCDR and qualified for command by the regulations of the parent Service is designated the commander of the Service component forces, unless another officer is so designated by competent authority. The assignment of any specific individual as a Service component commander of a CCMD is subject to the concurrence of the CCDR.

IV-15

Chapter IV

c. **Responsibilities of Service Component Commanders.** Service component commanders have responsibilities that derive from their roles in fulfilling the Services' support function. The JFC also may conduct operations through the Service component commander or, at lower echelons, other Service force commanders. In the event that a Service component commander exercises OPCON of forces and that OPCON over those forces is delegated by the JFC to another component commander or other subordinate JFC, the Service component commander retains the following responsibilities for certain Service-specific functions:

(1) Make recommendations to the JFC on the proper employment, task organization, and command relationship of the forces of the Service component.

(2) Accomplish such operational missions as may be assigned.

(3) Select and nominate specific units of the parent Service component for attachment to other subordinate commands. Unless otherwise directed, these units revert to the Service component commander's control when such subordinate commands are dissolved.

(4) Conduct joint training, including the training, as directed, of components of other Services in joint operations for which the Service component commander has or may be assigned primary responsibility, or for which the Service component's facilities and capabilities are suitable.

(5) Inform their JFC, other component or supporting commanders, and the CCDR, if affected, of planning for changes in logistic support that would significantly affect operational capability or sustainability early in the planning process for the JFC to evaluate the proposals prior to final decision or implementation. If the CCDR does not approve the proposal and discrepancies cannot be resolved between the JFC and the Service component commander, the CCDR will forward the issue through the CJCS to SecDef for resolution. Under crisis action or wartime conditions, and where critical situations make diversion of the normal logistic process necessary, Service component commanders will implement directives issued by the CCDR.

(6) Develop program and budget requests that comply with CCDR guidance on warfighting requirements and priorities. The Service component commander will provide to the CCDR a copy of the program submission prior to forwarding it to the Service headquarters. The Service component commander will keep the CCDR informed of the status of CCDR requirements while Service programs are under development.

(7) Inform the CCDR of program and budget decisions that may affect joint operation planning. The Service component commander will inform the CCDR of such decisions and of program and budget changes in a timely manner during the process in order to permit the CCDR to express the command's views before a final decision. The Service component commander will include in this information Service rationale for nonsupport of the CCDR's requirements.

IV-16

JP 1

Joint Command Organizations

(8) Provide, as requested, supporting joint operation and exercise plans with necessary force data to support missions that may be assigned by the CCDR.

d. Service component commanders or other Service force commanders assigned to a CCDR are responsible through the chain of command, extending to the Service Chief, for the following:

(1) Internal administration and discipline.

(2) Training in joint doctrine and their own Service doctrine, tactics, techniques, and procedures.

(3) Logistic functions normal to the command, except as otherwise directed by higher authority.

e. **Communication with a Service Chief.** Unless otherwise directed by the CCDR, the Service component commander will communicate through the CCMD on those matters over which the CCDR exercises COCOM. On Service-specific matters such as personnel, administration, and unit training, the Service component commander will normally communicate directly with the Service Chief, informing the CCDR as the CCDR directs.

f. **Logistic Authority.** The operating details of any Service logistic support system will be retained and exercised by the Service component commanders in accordance with instructions of their Military Departments, subject to the directive authority of the CCDR. Joint force transportation policies will comply with the guidelines established in the Defense Transportation System.

9. Functional Component Commands

a. JFCs have the authority to establish functional component commands to control military operations. JFCs may decide to establish a functional component command to integrate planning; reduce their span of control; and/or significantly improve combat efficiency, information flow, unity of effort, weapon systems management, component interaction, or control over the scheme of maneuver.

b. Functional component commanders have authority over forces or military capability made available to them as delegated by the establishing JFC. Functional component commands may be established to perform operational missions that may be of short or extended duration. JFCs may elect to centralize selected functions within the joint force, but should strive to avoid reducing the versatility, responsiveness, and initiative of subordinate forces. (Note: Functional component commanders are component commanders of a joint force and do not constitute a "joint force command" with the authorities and responsibilities of a JFC as described in this document, even when composed of forces from two or more Military Departments.)

c. The JFC establishing a functional component command has the authority to designate its commander. Normally, the Service component commander with the preponderance of forces to be tasked and the ability to C2 those forces will be designated

IV-17

Chapter IV

as the functional component commander; however, the JFC will always consider the mission, nature, and duration of the operation, force capabilities, and the C2 capabilities in selecting a commander.

d. The responsibilities and authority of a functional component command must be assigned by the establishing JFC. Establishment of a functional component command must not affect the command relationships between Service component and the JFC.

e. The JFC must designate the forces and/or military capability that will be made available for tasking by the functional component JFC and the appropriate command relationship(s) the functional component commander will exercise. Normally, the functional component commander will exercise OPCON over its own Service forces made available for tasking and TACON over other forces made available for tasking. The JFC may also establish support relationships between functional component commanders and other subordinate commanders to facilitate operations.

f. The commander of a functional component command is responsible for making recommendations to the establishing commander on the proper employment of the forces and/or military capability made available to accomplish the assigned responsibilities.

g. When a functional component command is composed of forces of two or more Services, the functional component commander must be cognizant of the constraints imposed by logistic factors on the capability of the forces attached or made available and the responsibilities retained by the Service component commanders.

h. When a functional component command will employ forces from more than one Service, the functional component JFC's staff should reflect the composition of the functional component command to provide the JFC with the expertise needed to effectively employ the forces made available. Staff billets for the needed expertise and individuals to fill those billets should be identified and used when the functional component staffs are formed for exercises and actual operations. The number of personnel on this staff should be kept to the minimum and should be consistent with the task performed. The structure of the staff should be flexible enough to expand or contract under changing conditions without a loss in coordination or capability.

For additional information on C2 by functional component commanders, see JP 3-30, Command and Control for Joint Air Operations; *JP 3-31,* Command and Control for Joint Land Operations; *JP 3-32,* Command and Control for Joint Maritime Operations; *and JP 3-05,* Special Operations.

SECTION C. DISCIPLINE

10. Responsibility

a. **Joint Force Commander.** The JFC is responsible for the discipline of military personnel assigned to the joint organization. In addition to the disciplinary authority exercised by subordinate JFCs, a CCDR may prescribe procedures by which the senior officer of a Service assigned to the headquarters element of a joint organization may exercise administrative and nonjudicial punishment authority over personnel of the same Service assigned to the same joint organization.

b. **Service Component Commander.** Each Service component in a CCMD is responsible for the discipline of that Service's component forces, subject to Service regulations and directives established by the CCDR. The JFC exercises disciplinary authority vested in the JFC by law, Service regulations, and superior authority in the chain of command.

c. **Method of Coordination.** The JFC normally exercises disciplinary authority through the Service component commanders to the extent practicable. When this is impracticable, the JFC may establish joint agencies responsible directly to the JFC to advise or make recommendations on matters placed within their jurisdiction or, if necessary, to carry out the directives of a superior authority. A joint military police force is an example of such an agency.

11. Uniform Code of Military Justice

The Uniform Code of Military Justice (UCMJ) is federal law, as enacted by Congress; it provides the basic law for discipline of the Armed Forces of the United States. The Manual for Courts-Martial (MCM), United States (as amended), prescribes the rules and procedures governing military justice. Pursuant to the authority vested in the President under article 22(a), UCMJ, and in Rules for Courts-Martial (RCM) 201(e)(2)(A) of the MCM (as amended), CCDRs are given courts-martial jurisdiction over members of any of the Armed Forces. Pursuant to article 23(a)(6), UCMJ, subordinate JFCs of a detached command or unit have special courts-martial convening authority. Under RCM 201(e)(2)(C), CCDRs may expressly authorize subordinate JFCs who are authorized to convene special and summary courts-martial to convene such courts-martial for the trial of members of other armed forces.

12. Rules and Regulations

Rules and regulations implementing the UCMJ and MCM are, for the most part, of single-Service origin. In a joint force, however, the JFC should publish rules and regulations that establish uniform policies applicable to all Services' personnel within the joint organization where appropriate. For example, joint rules and regulations should be published to address hours and areas authorized for liberty, apprehension of Service personnel, black market activities, prescription drug and alcohol consumption, combating trafficking in persons, sexual assault prevention and response policies, currency control regulations, and any other matters that the JFC deems appropriate.

13. Jurisdiction

a. **More Than One Service Involved.** Matters that involve more than one Service and that are within the jurisdiction of the JFC may be handled either by the JFC or by the appropriate Service component commander.

Chapter IV

b. **One Service Involved.** Matters that involve only one Service should be handled by the Service component commander, subject to Service regulations. A Service member is vested with a hierarchy of rights. From greatest to least, these are: the US Constitution, the UCMJ, departmental regulations, Service regulations, and the common law. JFCs must ensure that an accused Service member's rights are not abridged or trampled. National Guard members operating under Title 32, USC, or state active duty status are subject to their respective state's jurisdiction.

c. A commander's UCMJ jurisdiction may extend over persons serving with or accompanying the Armed Forces of the United States during military operations in accordance with UCMJ Article 2 (Title 10, USC, Section 802). Criminal matters falling outside UCMJ jurisdiction may be managed under Title 18, USC, Sections 3261–3267 (also known as the Military Extraterritorial Jurisdiction Act).

For further information on appropriate procedures for exercising criminal jurisdiction over civilians employed by or accompanying the force, see DODI 5525.11, Criminal Jurisdiction Over Civilians Employed By or Accompanying the Armed Forces Outside the United States, Certain Service Members, and Former Service Members.

14. Trial and Punishment

a. **Convening Courts-Martial.** General courts-martial may be convened by the CCDR. An accused may be tried by any courts-martial convened by a member of a different Service when the courts-martial is convened by a JFC who has been specifically empowered by statute, the President, SecDef, or a superior commander under the provisions of the RCM (201[e][2] of the MCM) to refer such cases for trial by courts-martial.

b. **Post-Trial and Appellate Processing.** When a court-martial is convened by a JFC, the convening authority may take action on the sentence and the findings as authorized by the UCMJ and MCM. If the convening authority is unable to take action, the case will be forwarded to an officer exercising general court-martial jurisdiction. Following convening authority action, the review and appeals procedures applicable to the accused's Service will be followed.

c. **Nonjudicial Punishment.** The JFC may impose nonjudicial punishment upon any military personnel of the command, unless such authority is limited or withheld by a superior commander. The JFC will use the regulations of the alleged offender's Service when conducting nonjudicial punishment proceedings, including punishment, suspension, mitigation, and filing. However, the JFC should normally allow nonjudicial punishment authority to be exercised by the appropriate Service component commander. Except as noted below, appeals and other actions involving review of nonjudicial punishment imposed by a JFC will follow the appropriate regulations of the alleged offender's Service. When the CCDR personally imposes nonjudicial punishment, or is otherwise disqualified from being the appellate authority, appeals will be forwarded to the CJCS for appropriate action by SecDef or SecDef designee. Collateral decisions and processing (e.g., personnel

Joint Command Organizations

and finance actions and unfavorable notations in selection records and personnel files) will be handled in Service channels.

d. **Execution of Punishment.** Execution of any punishment adjudged or imposed within any Service may be carried out by another Service under regulations provided by the Secretaries of the Military Departments.

SECTION D. PERSONNEL SERVICE SUPPORT AND ADMINISTRATION

15. Morale, Welfare, and Recreation

In a joint force, the morale and welfare of each Service member is the responsibility of the Service component commander. The JFC coordinates morale, welfare, and recreation (MWR) programs within the operational area. MWR facilities may be operated either by a single Service or jointly as directed by the CCDR in whose AOR the facility is located. Facilities operated by one Service should be made available to personnel of other Services to the extent practicable. Facilities that are jointly operated should be available equitably to all Services.

For further information on MWR, see JP 1-0, Joint Personnel Support.

16. Awards and Decorations

Recommendations for decorations and medals will be made by the JFC in accordance with Service regulations or as prescribed by DOD 1348.33, *Manual of Military Decorations and Awards, Volumes I and II,* as applicable. Recommendations for joint awards will be processed through joint channels. Concurrence of the CJCS is required prior to initiating a request for a joint award for a CCDR. When a member of a joint staff is recommended for a Service award, the JFC will process the recommendation through Service channels. Forward offers of personal foreign decorations through the CCDR to the Secretary of the appropriate Military Department. Forward offers of foreign unit, Service, or campaign medals through the CCDR to the CJCS.

17. Efficiency, Fitness, and Performance Reports

The immediate superior of an officer or enlisted Service member in a joint organization is responsible for preparing an efficiency, fitness, or performance report in accordance with the guidance (and on the prescribed form) of the rated member's Service. Completed reports will be forwarded to the reported Service member's Service for filing. A copy of the signed report will be provided to the Service member, and a copy will be maintained by the senior rater in accordance with Service guidance. Letter reports prepared by CCDRs for component commanders will be forwarded through the CJCS to the reported officer's Service.

For further information on personnel administration, refer to JP 1-0, Joint Personnel Support.

IV-21

Chapter IV

18. Total Force Fitness

a. The most valuable resource in our military—individual Service members—are continually confronted with considerable, sustained, and diverse stressors that not only impact their health and well-being, but impact the health and well-being of their family and may ultimately impact the ability of their unit and the US Armed Forces to accomplish assigned missions.

b. Total force fitness (TFF) provides an integrative and holistic framework to better understand, assess, and maintain the fitness of the joint force. It assists individuals in sustaining their well-being, which directly contributes to their ability to carry out assigned tasks. Leaders at all levels must understand, establish, and support a TFF program within their organizations.

c. The character, professionalism, and values that are the hallmark of the joint force demand that leaders care for our forces and their families. TFF offers a model for meeting that demand.

For additional details on TFF, see CJCSI 3405.01, Chairman's Total Force Fitness Framework.

19. Personnel Accountability

The JFC will establish standardized procedures to account for all personnel composing the force to include obtaining initial accountability and continuous updates throughout the duration of the operation. The JFC accomplishes joint personnel strength reporting and manages casualty reporting. The JFC provides personnel reports to the CCDR and CJCS as directed.

20. Religious Affairs

Religious affairs are the commander's responsibility and consist of the combination of religious support and religious advisement. Religious support addresses the joint commander's responsibilities to support the free exercise of religion by members of the joint force to the standards set by DOD and the Services and to make a good faith effort in support of the welfare of personnel. Religious advisement addresses the commander's requirement to receive germane subject matter advice on the impact of religion on operations. All military commanders are responsible for religious affairs in their command. While the Services set standards for their personnel, each commander is responsible for identifying religious requirements unique to his or her echelon and circumstances. Religious support consists of the accommodation of the free exercise of religious beliefs through provision and facilitation of religious worship and pastoral care; advising the JFC on ethics, morals, morale, command climate, and the command religious program. Chaplains work to fulfill religious requirements in coordination with other chaplains and with the aid of chaplain's assistants and religious program specialists. Chaplain confidentiality ensures that all members of the joint force, regardless of religious identity, have the opportunity to seek human care from professionals who cannot disclose the content of communications. Religious affairs are conducted according to Service policies and standards.

For further information on religious support, see JP 1-05, Religious Affairs in Joint Operations.

21. Information Management

The JFC should ensure that all information is treated as record material and properly handled to meet statutory requirements and sound records management principles as defined in DODD 5015.2, *DOD Records Management Program,* and supported by CJCSI 5760.01, *Records Management Policy for the Joint Staff and Combatant Commands.*

Chapter IV

Intentionally Blank

CHAPTER V
JOINT COMMAND AND CONTROL

"[My job is] to give the President and the Secretary of Defense military advice before they know they need it."

**General John W. Vessey, Jr., US Army
Chairman of the Joint Chiefs of Staff
(18 June 1982–30 September 1985)**

SECTION A. COMMAND RELATIONSHIPS

1. General Principles

a. **Command.** Command is central to all military action, and unity of command is central to unity of effort. Inherent in command is the authority that a military commander lawfully exercises over subordinates including authority to assign missions and accountability for their successful completion. **Although commanders may delegate authority to accomplish missions, they may not absolve themselves of the responsibility for the attainment of these missions.** Authority is never absolute; the extent of authority is specified by the establishing authority, directives, and law.

b. **Unity of Command and Unity of Effort.** Unity of command means all forces operate under a single commander with the requisite authority to direct all forces employed in pursuit of a common purpose. Unity of effort, however, requires coordination and cooperation among all forces toward a commonly recognized objective, although they are not necessarily part of the same command structure. During multinational operations and interagency coordination, unity of command may not be possible, but the requirement for unity of effort becomes paramount. Unity of effort—coordination through cooperation and common interests—is an essential complement to unity of command. **Unity of command requires that two commanders may not exercise the same command relationship over the same force at any one time.**

c. **Command and Staff.** JFCs are provided staffs to assist them in the decision-making and execution process. The staff is an extension of the JFC; its function is command support and its authority is delegated by the JFC. A properly trained and directed staff will free the JFC to devote more attention to directing subordinate commanders and maintaining a picture of the overall situation.

(1) Chain of command is the succession of commanding officers from a superior to a subordinate through which command is exercised.

(2) Staffing is the term used to describe the coordination between staffs at higher, adjacent, and subordinate headquarters. Higher headquarters staff officers exercise no independent authority over subordinate headquarters staffs, although staff officers normally respond to requests for information.

V-1

Chapter V

d. **Levels of Authority.** The specific command relationship (COCOM, OPCON, TACON, and support) will define the authority a commander has over assigned or attached forces. An overview of command relationships is shown in Figure V-1.

2. Combatant Command (Command Authority)

COCOM is the command authority over assigned forces vested only in the commanders of CCMDs by Title 10, USC, Section 164 (or as otherwise directed by the President or SecDef) and cannot be delegated or transferred.

a. **Basic Authority.** COCOM provides full authority for a CCDR to perform those functions of command over assigned forces involving organizing and employing commands and forces, assigning tasks, designating objectives, and giving authoritative direction over all aspects of military operations, joint training (or in the case of USSOCOM, training of assigned forces), and logistics necessary to accomplish the missions assigned to the command. COCOM should be exercised through the commanders of subordinate organizations, normally JFCs, Service and/or functional component commanders.

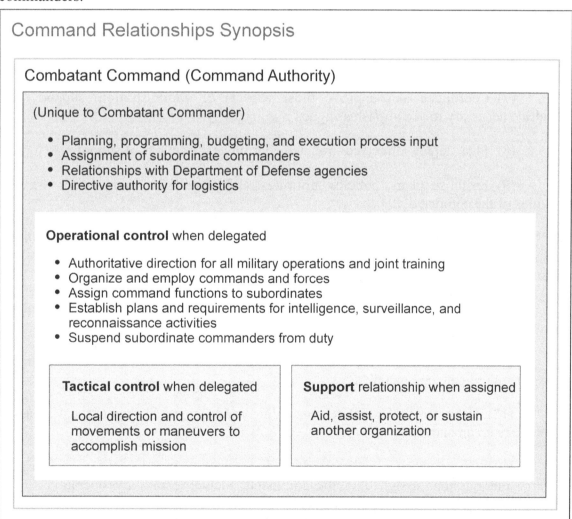

Figure V-1. Command Relationships Synopsis

b. Unless otherwise directed by the President or SecDef, the authority, direction, and control of the CCDR with respect to the command of forces assigned to that command includes the following:

(1) Exercise or delegate OPCON, TACON, or other specific elements of authority, establish support relationships among subordinate commanders over assigned or attached forces, and designate coordinating authorities, as described below.

(2) Exercise directive authority for logistic matters (or delegate directive authority for a common support capability to a subordinate commander via an establishing directive).

(3) Prescribe the chain of command to the commands and forces within the command.

(4) Organize subordinate commands and forces within the command as necessary to carry out missions assigned to the command.

(5) Employ forces within that command as necessary to carry out missions assigned to the command.

(6) Assign command functions to subordinate commanders.

(7) Coordinate and approve those aspects of administration, support, and discipline necessary to accomplish assigned missions.

(8) Plan, deploy, direct, control, and coordinate the actions of subordinate forces.

(9) Establish plans, policies, priorities, and overall requirements for the ISR activities of the command.

(10) Conduct joint exercises and training to achieve effective employment of the forces in accordance with joint and established training policies for joint operations. This authority also applies to forces attached for purposes of joint exercises and training.

(11) Assign responsibilities to subordinate commanders for certain routine operational matters that require coordination of effort of two or more commanders.

(12) Establish a system of control for local defense and delineate such operational areas for subordinate commanders.

(13) Delineate functional responsibilities and geographic operational areas of subordinate commanders.

(14) Give authoritative direction to subordinate commands and forces necessary to carry out missions assigned to the command, including military operations, joint training, and logistics.

Chapter V

(15) Coordinate with other CCDRs, USG departments and agencies, and organizations of other countries regarding matters that cross the boundaries of geographic areas specified in the UCP and inform USG departments and agencies or organizations of other countries in the AOR, as necessary, to prevent both duplication of effort and lack of adequate control of operations in the delineated areas.

(16) Unless otherwise directed by SecDef, function as the US military single point of contact and exercise directive authority over all elements of the command in relationships with other CCMDs, DOD elements, US diplomatic missions, other USG departments and agencies, and organizations of other countries in the AOR. Whenever a CCDR conducts exercises, operations, or other activities with the military forces of nations in another CCDR's AOR, those exercises, operations, and activities and their attendant command relationships will be mutually agreed to between the CCDRs.

(17) Determine those matters relating to the exercise of COCOM in which subordinates must communicate with agencies external to the CCMD through the CCDR.

(18) Establish personnel policies to ensure proper and uniform standards of military conduct.

(19) Submit recommendations through the CJCS to SecDef concerning the content of guidance affecting the strategy and/or fielding of joint forces.

(20) Participate in the planning, programming, budgeting, and execution process as specified in appropriate DOD issuances.

(21) Participate in the JSPS and the APEX system. CCDRs' comments are critical to ensuring that warfighting and peacetime operational concerns are emphasized in all planning documents.

(22) Concur in the assignment (or recommendation for assignment) of officers as commanders directly subordinate to the CCDR and to positions on the CCMD staff. Suspend from duty and recommend reassignment, when appropriate, any subordinate officer assigned to the CCMD.

(23) Convene general courts-martial in accordance with the UCMJ.

(24) In accordance with laws and national and DOD policies, establish plans, policies, programs, priorities, and overall requirements for the C2, communications system, and ISR activities of the command.

c. When directed in the UCP or otherwise authorized by SecDef, the commander of US elements of a multinational command may exercise COCOM of those US forces assigned to that command (e.g., United States Element, North American Aerospace Defense Command [USELEMNORAD]).

d. **Directive Authority for Logistics.** CCDRs exercise directive authority for logistics and may delegate directive authority for a common support capability to a

Joint Command and Control

subordinate JFC as required to accomplish the subordinate JFC's assigned mission. For some commodities or support services common to two or more Services, one Service may be given responsibility for management based on DOD EA designations or inter-Service support agreements. However, the CCDR must formally delineate this delegated directive authority by function and scope to the subordinate JFC or Service component commander. The exercise of directive authority for logistics by a CCDR includes the authority to issue directives to subordinate commanders, including peacetime measures necessary to ensure the following: effective execution of approved plans, effectiveness and economy of operation, and prevention or elimination of unnecessary duplication of facilities and overlapping of functions among the Service component commands. CCDRs will coordinate with appropriate Services before exercising directive authority for logistics or delegate authority for subordinate commanders to exercise common support capabilities to one of their components.

(1) A CCDR's directive authority does not:

(a) Discontinue Service responsibility for logistic support.

(b) Discourage coordination by consultation and agreement.

(c) Disrupt effective procedures or efficient use of facilities or organizations.

(d) Include the ability to provide contracting authority or make binding contracts for the USG.

(2) Unless otherwise directed by SecDef, the Military Departments and Services continue to have responsibility for the logistic support of their forces assigned or attached to joint commands, subject to the following guidance:

(a) Under peacetime conditions, the scope of the logistic authority exercised by the commander of a CCMD will be consistent with the peacetime limitations imposed by legislation, DOD policy or regulations, budgetary considerations, local conditions, and other specific conditions prescribed by SecDef or the CJCS. Where these factors preclude execution of a CCDR's directive by component commanders, the comments and recommendations of the CCDR, together with the comments of the component commander concerned, normally will be referred to the appropriate Military Department for consideration. If the matter is not resolved in a timely manner with the appropriate Military Department, it will be referred by the CCDR, through the CJCS, to SecDef.

(b) Under crisis action, wartime conditions, or where critical situations make diversion of the normal logistic process necessary, the logistic authority of CCDRs enables them to use all facilities and supplies of all forces assigned to their commands for the accomplishment of their missions. The President or SecDef may extend this authority to attached forces when transferring those forces for a specific mission and should specify this authority in the establishing directive or order. Joint logistic doctrine and policy developed by the CJCS establishes wartime logistic support guidance to assist the CCDR in conducting successful joint operations.

Chapter V

For further information on logistic support, refer to JP 4-0, Joint Logistics.

(3) A CCDR will exercise approval authority over Service logistic programs (base adjustments, force basing, and other aspects, as appropriate) within the command's AOR that will have a significant impact on operational capability or sustainability.

3. Operational Control

OPCON is the command authority that may be exercised by commanders at any echelon at or below the level of CCMD and may be delegated within the command.

a. **Basic Authority.** OPCON is able to be delegated from a lesser authority than COCOM. It is the authority to perform those functions of command over subordinate forces involving organizing and employing commands and forces, assigning tasks, designating objectives, and giving authoritative direction over all aspects of military operations and joint training necessary to accomplish the mission. It should be delegated to and exercised by the commanders of subordinate organizations; normally, this authority is exercised through subordinate JFCs, Service, and/or functional component commanders. OPCON provides authority to organize and employ commands and forces as the commander considers necessary to accomplish assigned missions. It does not include authoritative direction for logistics or matters of administration, discipline, internal organization, or unit training. These elements of COCOM must be specifically delegated by the CCDR. OPCON does include the authority to delineate functional responsibilities and operational areas of subordinate JFCs.

b. Commanders of subordinate commands, including JTFs, will be given OPCON of assigned forces and OPCON or TACON of attached forces by the superior commander.

c. OPCON includes the authority for the following:

(1) Exercise or delegate OPCON and TACON or other specific elements of authority and establish support relationships among subordinates, and designate coordinating authorities.

(2) Give direction to subordinate commands and forces necessary to carry out missions assigned to the command, including authoritative direction over all aspects of military operations and joint training.

(3) Prescribe the chain of command to the commands and forces within the command.

(4) With due consideration for unique Service organizational structures and their specific support requirements, organize subordinate commands and forces within the command as necessary to carry out missions assigned to the command.

(5) Employ forces within the command, as necessary, to carry out missions assigned to the command.

V-6   JP 1

Joint Command and Control

(6) Assign command functions to subordinate commanders.

(7) Plan for, deploy, direct, control, and coordinate the actions of subordinate forces.

(8) Establish plans, policies, priorities, and overall requirements for the ISR activities of the command.

(9) Conduct joint training exercises required to achieve effective employment of the forces of the command, in accordance with joint doctrine established by the CJCS, and establish training policies for joint operations required to accomplish the mission. This authority also applies to forces attached for purposes of joint exercises and training.

(10) Suspend from duty and recommend reassignment of any officer assigned to the command.

(11) Assign responsibilities to subordinate commanders for certain routine operational matters that require coordination of effort of two or more commanders.

(12) Establish an adequate system of control for local defense and delineate such operational areas for subordinate commanders as deemed desirable.

(13) Delineate functional responsibilities and geographic operational areas of subordinate commanders.

d. SecDef may specify adjustments to accommodate authorities beyond OPCON in an establishing directive when forces are transferred between CCDRs or when members and/or organizations are transferred from the Military Departments to a CCMD. Adjustments will be coordinated with the participating Service Chiefs and CCDRs.

4. Tactical Control

TACON is an authority over assigned or attached forces or commands, or military capability or forces made available for tasking, that is limited to the detailed direction and control of movements and maneuvers within the operational area necessary to accomplish assigned missions or tasks assigned by the commander exercising OPCON or TACON of the attached force.

a. **Basic Authority.** TACON is able to be delegated from a lesser authority than OPCON and may be delegated to and exercised by commanders at any echelon at or below the level of CCMD.

b. **TACON provides the authority to:**

(1) Give direction for military operations; and

(2) Control designated forces (e.g., ground forces, aircraft sorties, or missile launches).

V-7

Chapter V

c. TACON does not provide the authority to give or change the function of the subordinate commander.

d. TACON provides sufficient authority for controlling and directing the application of force or tactical use of combat support assets within the assigned mission or task. **TACON does not provide organizational authority or authoritative direction for administrative and logistic support.**

e. Functional component commanders typically exercise TACON over military capability or forces made available for tasking.

5. Support

Support is a command authority. A support relationship is established by a common superior commander between subordinate commanders when one organization should aid, protect, complement, or sustain another force. The support command relationship is used by SecDef to establish and prioritize support between and among CCDRs, and it is used by JFCs to establish support relationships between and among subordinate commanders.

a. **Basic Authority.** Support may be exercised by commanders at any echelon at or below the CCMD level. The designation of supporting relationships is important as it conveys priorities to commanders and staffs that are planning or executing joint operations. The support command relationship is, by design, a somewhat vague but very flexible arrangement. The establishing authority (the common JFC) is responsible for ensuring that both the supported commander and supporting commanders understand the degree of authority that the supported commander is granted.

b. The supported commander should ensure that the supporting commanders understand the assistance required. The supporting commanders will then provide the assistance needed, subject to a supporting commander's existing capabilities and other assigned tasks. When a supporting commander cannot fulfill the needs of the supported commander, the establishing authority will be notified by either the supported commander or a supporting commander. The establishing authority is responsible for determining a solution.

c. An establishing directive normally is issued to specify the purpose of the support relationship, the effect desired, and the scope of the action to be taken. It also should include:

(1) The forces and resources allocated to the supporting effort;

(2) The time, place, level, and duration of the supporting effort;

(3) The relative priority of the supporting effort;

(4) The authority, if any, of the supporting commander to modify the supporting effort in the event of exceptional opportunity or an emergency; and

V-8   JP 1

Joint Command and Control

(5) The degree of authority granted to the supported commander over the supporting effort.

d. Unless limited by the establishing directive, the supported commander will have the authority to exercise general direction of the supporting effort. General direction includes the designation and prioritization of targets or objectives, timing and duration of the supporting action, and other instructions necessary for coordination and efficiency.

e. The supporting commander determines the forces, tactics, methods, procedures, and communications to be employed in providing this support. The supporting commander will advise and coordinate with the supported commander on matters concerning the employment and limitations (e.g., sustainment) of such support, assist in planning for the integration of such support into the supported commander's effort as a whole, and ensure that support requirements are appropriately communicated within the supporting commander's organization.

f. The supporting commander has the responsibility to ascertain the needs of the supported force and take action to fulfill them within existing capabilities, consistent with priorities and requirements assigned tasks.

g. There are four categories of support that a CCDR may exercise over assigned or attached forces to ensure the appropriate level of support is provided to accomplish mission objectives. They are: general support, mutual support, direct support, and close support. For example, land forces that provide fires normally are tasked in a direct support role. Figure V-2 summarizes each of the categories of support. The establishing directive will specify the type and extent of support the specified forces are to provide.

For further information, see Appendix A, "Establishing Directive (Support Relationship) Considerations."

6. Support Relationships Between Combatant Commanders

a. SecDef establishes support relationships between the CCDRs for the planning and execution of joint operations. This ensures that the supported CCDR receives the necessary support. A supported CCDR requests capabilities, tasks supporting DOD components, coordinates with the appropriate USG departments and agencies (where agreements have been established), and develops a plan to achieve the common goal. As part of the team effort, supporting CCDRs provide the requested capabilities, as available, to assist the supported CCDR to accomplish missions requiring additional resources.

b. The CJCS organizes the JPEC for joint operation planning to carry out support relationships between the CCMDs. The supported CCDR has primary responsibility for all aspects of an assigned task. Supporting CCDRs provide forces, assistance, or other resources to a supported CCDR. Supporting CCDRs prepare supporting plans as required. Under some circumstances, a CCDR may be a supporting CCDR for one operation while being a supported CCDR for another.

V-9

Chapter V

Categories of Support

General Support

That support that is given to the supported force as a whole rather than to a particular subdivision thereof.

Mutual Support

That support that units render each other against an enemy because of their assigned tasks, their position relative to each other and to the enemy, and their inherent capabilities.

Direct Support

A mission requiring a force to support another specific force and authorizing it to answer directly to the supported force's request for assistance.

Close Support

That action of the supporting force against targets or objectives that are sufficiently near the supported force as to require detailed integration or coordination of the supporting action with the fire, movement, or other actions of the supported force.

Figure V-2. Categories of Support

7. Support Relationships Between Component Commanders

a. The JFC may establish support relationships between component commanders to facilitate operations. Support relationships afford an effective means to prioritize and ensure unity of effort for various operations. Component commanders should establish liaison with other component commanders to facilitate the support relationship and to coordinate the planning and execution of pertinent operations. Support relationships may change across phases of an operation as directed by the establishing authority.

b. When the commander of a Service component is designated as a functional component commander, the associated Service component responsibilities for assigned or attached forces are retained, but are not applicable to forces made available by other Service components. The operational requirements of the functional component commander's subordinate forces are prioritized and presented to the JFC by the functional component commander, relieving the affected Service component commanders of this responsibility, but the affected Service component commanders are not relieved of their administrative and support responsibilities.

c. In rare situations, a supporting component commander may be supporting two or more supported commanders. In these situations, there must be clear understanding among all parties, and a specification in the establishing directive, as to who supports whom, when, and with what prioritization. When there is a conflict over prioritization, employment, or task organization between component commanders, the CCDR having COCOM of the component commanders will adjudicate.

V-10

JP 1

Joint Command and Control

8. Command Relationships and Assignment and Transfer of Forces

All forces under the jurisdiction of the Secretaries of the Military Departments (except those forces necessary to carry out the functions of the Military Departments as noted in Title 10, USC, Section 162) are assigned to CCMDs or Commander, USELEMNORAD, or designated as Service retained by SecDef in the GFMIG. A force assigned or attached to a CCMD, or Service retained by a Service Secretary, may be transferred from that command to another CCDR only when directed by SecDef and under procedures prescribed by SecDef and approved by the President. The command relationship the gaining commander will exercise (and the losing commander will relinquish) will be specified by SecDef. Establishing authorities for subordinate unified commands and JTFs may direct the assignment or attachment of their forces to those subordinate commands and delegate the command relationship as appropriate (see Figure V-3).

a. The CCDR exercises COCOM over forces assigned or reassigned by the President or SecDef. Forces are assigned or reassigned when the transfer of forces will be permanent or for an unknown period of time, or when the broadest command authority is required or desired. OPCON of assigned forces is inherent in COCOM and may be delegated within the CCMD by the CCDR.

b. The CCDR normally exercises OPCON over forces attached by SecDef. Forces are attached when the transfer of forces will be temporary. Establishing authorities for subordinate unified commands and JTFs normally will direct the delegation of OPCON over forces attached to those subordinate commands.

c. In accordance with the GFMIG and the UCP, except as otherwise directed by the President or SecDef, all forces operating within the geographic area assigned to a specific GCC shall be assigned or attached to, and under the command of, that GCC. (This does

---

Transfer of Forces and Command Relationships Overview

- Forces, not command relationships, are transferred between commands. When forces are transferred, the command relationship the gaining commander will exercise (and the losing commander will relinquish) over those forces must be specified.

- When transfer of forces to a joint force will be permanent (or for an unknown but long period of time) the forces should be reassigned. Combatant commanders will exercise combatant command (command authority), and subordinate joint force commanders (JFCs), normally through the Service component commander, will exercise operational control (OPCON) over reassigned forces.

- When transfer of forces to a joint force will be temporary, the forces will be attached to the gaining command, and JFCs, normally through the Service component commander, will exercise OPCON over the attached forces.

- Establishing authorities for subordinate unified commands and joint task forces direct the assignment or attachment of their forces to those subordinate commands as appropriate.

Figure V-3. Transfer of Forces and Command Relationships Overview

Chapter V

not apply to USNORTHCOM.) Transient forces do not come under the chain of command of the GCC solely by their movement across operational area boundaries, except when the GCC is exercising TACON for the purpose of force protection. Unless otherwise specified by SecDef, and with the exception of the USNORTHCOM AOR, a GCC has TACON for exercise purposes whenever forces not assigned to that GCC undertake exercises in that GCC's AOR.

9. Other Authorities

Other authorities outside the command relationships delineated above are described below.

a. **Administrative Control.** ADCON is the direction or exercise of authority over subordinate or other organizations with respect to administration and support, including organization of Service forces, control of resources and equipment, personnel management, logistics, individual and unit training, readiness, mobilization, demobilization, discipline, and other matters not included in the operational missions of the subordinate or other organizations. ADCON is synonymous with administration and support responsibilities identified in Title 10, USC. This is the authority necessary to fulfill Military Department statutory responsibilities for administration and support. ADCON may be delegated to and exercised by commanders of Service forces assigned to a CCDR at any echelon at or below the level of Service component command. ADCON is subject to the command authority of CCDRs. ADCON may be delegated to and exercised by commanders of Service commands assigned within Service authorities. Service commanders exercising ADCON will not usurp the authorities assigned by a CCDR having COCOM over commanders of assigned Service forces.

b. **Coordinating Authority.** Commanders or individuals may exercise coordinating authority at any echelon at or below the level of CCMD. Coordinating authority is the authority delegated to a commander or individual for coordinating specific functions and activities involving forces of two or more Military Departments, two or more joint force components, or two or more forces of the same Service (e.g., joint security coordinator exercises coordinating authority for joint security area operations among the component commanders). Coordinating authority may be granted and modified through an MOA to provide unity of effort for operations involving RC and AC forces engaged in interagency activities. The commander or individual has the authority to require consultation between the agencies involved but does not have the authority to compel agreement. The common task to be coordinated will be specified in the establishing directive without disturbing the normal organizational relationships in other matters. Coordinating authority is a consultation relationship between commanders, not an authority by which command may be exercised. It is more applicable to planning and similar activities than to operations. Coordinating authority is not in any way tied to force assignment. Assignment of coordinating authority is based on the missions and capabilities of the commands or organizations involved.

c. **Direct Liaison Authorized (DIRLAUTH).** DIRLAUTH is that authority granted by a commander (any level) to a subordinate to directly consult or coordinate an action

V-12   JP 1

Joint Command and Control

with a command or agency within or outside of the granting command. DIRLAUTH is more applicable to planning than operations and always carries with it the requirement of keeping the commander granting DIRLAUTH informed. DIRLAUTH is a coordination relationship, not an authority through which command may be exercised.

10. Command of National Guard and Reserve Forces

a. When mobilized under Title 10, USC, authority, command of National Guard and Reserve forces (except those forces specifically exempted) is assigned by SecDef to the CCMDs. Those forces are available for operational missions when mobilized for specific periods or when ordered to active duty after being validated for employment by their parent Service. Normally, National Guard forces are under the commands of their respective governors in Title 32, USC, or state active duty status.

b. The authority CCDRs may exercise over assigned RC forces when not on active duty or when on active duty for training is TRO. CCDRs normally will exercise TRO over assigned forces through Service component commanders. TRO includes authority to:

(1) Provide guidance to Service component commanders on operational requirements and priorities to be addressed in Military Department training and readiness programs;

(2) Comment on Service component program recommendations and budget requests;

(3) Coordinate and approve participation by assigned RC forces in joint exercises and training when on active duty for training or performing inactive duty for training;

(4) Obtain and review readiness and inspection reports on assigned RC forces; and

(5) Coordinate and review mobilization plans (including post-mobilization training and deployability validation) developed for assigned RC forces.

c. Unless otherwise directed by SecDef, the following applies:

(1) Assigned RC forces on active duty (other than for training) may not be deployed until validated by the parent Service for deployment.

(2) CDRs may employ RC forces assigned to subordinate component commanders in contingency operations when forces have been mobilized for specific periods or when ordered to active duty after being validated for employment by their parent Service.

(3) RC forces on active duty for training or performing inactive-duty training may be employed in connection with contingency operations only as provided by law, and when the primary purpose is for training consistent with their mission or specialty.

Chapter V

d. CCDRs will communicate with assigned RC forces through the Military Departments when the RC forces are not on active duty or when on active duty for training.

e. CCDRs may inspect assigned RC forces in accordance with DODD 5106.04, *Combatant Command Inspectors General,* when such forces are mobilized or ordered to active duty (other than for training).

f. CDRUSSOCOM will exercise additional authority for certain functions for assigned RC forces and for all SOF assigned to other CCMDs in accordance with the current MOAs between CDRUSSOCOM and the Secretaries of the Military Departments. See DODI 1215.06, *Uniform Reserve, Training, and Retirement Categories.*

SECTION B. COMMAND AND CONTROL OF JOINT FORCES

11. Background

Command is the most important role undertaken by a JFC. It is the exercise of authority and direction by a properly designated commander over assigned and attached forces. C2 is the means by which a JFC synchronizes and/or integrates joint force activities. C2 ties together all the operational functions and tasks and applies to all levels of war and echelons of command. C2 functions are performed through an arrangement of personnel, equipment, communications, facilities, and procedures employed by a commander in planning, directing, coordinating, and controlling forces and operations in the accomplishment of the mission.

12. Command and Control Fundamentals

C2 enhances the commander's ability to make sound and timely decisions and successfully execute them. Unity of effort over complex operations is made possible through decentralized execution of centralized, overarching plans or via mission command. Advances in information systems and communications may enhance the situational awareness (SA) and understanding of tactical commanders, subordinate JFCs, CCDRs, and even the national leadership. The level of control used will depend on the nature of the operation or task, the risk or priority of its success, and the associated comfort level of the commander.

a. **Tenets.** Unity of command is strengthened through adherence to the following C2 tenets:

(1) Clearly Defined Authorities, Roles, and Relationships. Effective C2 of joint operations begins by establishing unity of command through the designation of a JFC with the requisite authority to accomplish assigned tasks using an uncomplicated chain of command. It is essential for the JFC to ensure that subordinate commanders, staff principals, and leaders of C2 nodes (e.g., IO cell, joint movement center) understand their authorities, role in decision making and controlling, and relationships with others. The assignment of responsibilities and the delegation of authorities foster initiative and speed the C2 process. Joint force staff principals must understand that their primary role is to provide sufficient, relevant information to enhance SA and understanding for the JFC and

Joint Command and Control

for subordinate commanders. Once a decision is made, commanders depend on their staffs to communicate the decision to subordinates in a manner that quickly focuses the necessary capabilities within the command to achieve the commander's intent. The commander should give the staff the authority to make routine decisions within the constraints of the commander's intent while conducting operations. Appropriate application of the command relationships discussed previously in Section A will help ensure that the requisite amount of control is applied while enabling sufficient latitude for decentralized execution. Additionally, commander to staff and staff to staff relationships must be developed through training to promote the understanding of all regarding the direction and/or support required.

(2) **Mission command** is the conduct of military operations through decentralized execution based upon mission-type orders. It empowers individuals to exercise judgment in how they carry out their assigned tasks and it exploits the human element in joint operations, emphasizing trust, force of will, initiative, judgment, and creativity. Successful mission command demands that subordinate leaders at all echelons exercise disciplined initiative and act aggressively and independently to accomplish the mission. They focus their orders on the purpose of the operation rather than on the details of how to perform assigned tasks. They delegate decisions to subordinates wherever possible, which minimizes detailed control and empowers subordinates' initiative to make decisions based on understanding what the commander wants rather than on constant communications. Essential to mission command is the thorough understanding of the *commander's intent* at every level of command and a command climate of mutual trust and understanding.

(3) Information Management and Knowledge Sharing. Control and appropriate sharing of information is a prerequisite to maintaining effective C2. For a discussion of information management and knowledge sharing, see JP 3-0, *Joint Operations*, and JP 3-33, *Joint Task Force Headquarters.*

(4) Communication. Because JFCs seek to minimize restrictive control measures and detailed instructions, they must find effective and efficient ways to create cooperation and compliance. Commander's intent fosters communication and understanding with all subordinates. This common understanding builds teamwork and mutual trust. Two joint C2 constructs that ensure implicit communication are the commander's intent and mission statement.

(a) Commander's intent represents a unifying idea that allows decentralized execution within centralized, overarching guidance. It is a clear and concise expression of the purpose of the operation and the military end state. It provides focus to the staff and helps subordinate and supporting commanders take actions to achieve the military end state without further orders, even when operations do not unfold as planned.

(b) JFCs use mission-type orders to decentralize execution. Mission-type orders direct a subordinate to perform a certain task without specifying how to accomplish it. Within these orders, the actual mission statement should be a short sentence or paragraph that describes the organization's essential task (or tasks) and purpose—a clear statement of the action to be taken and the reason for doing so. The senior leaves the details

V-15

Chapter V

of execution to the subordinate, allowing the freedom and the obligation to take whatever steps are necessary to deal with the changing situation while encouraging initiative at lower levels.

(5) Timely Decision Making. With well-defined commander's critical information requirement, effective common operational picture and establishing clear objectives, the JFC can make timely and effective decisions to get inside the adversary's decision and execution cycle. Doing so generates confusion and disorder and slows an adversary's decision making. The commander who can gather information and make better decisions faster will generate a rapid tempo of operations and gain a decided advantage. Consequently, decision-making models and procedures must be flexible and allow abbreviation should the situation warrant it. Adoption of a decision aid offers the commander and staff a method for maintaining SA of the ongoing operation as well as identifying critical decision points where the commander's action may be required to maintain momentum.

(6) Coordination Mechanisms. Coordination mechanisms facilitate integration, synchronization, and synergistic interaction among joint force components. Coordinating mechanisms can include: agreements, memoranda of understanding, exchange and/or liaison officers, direct and integrated staffing, interoperable communications systems, information sharing, exercises, and plan development. Integration is achieved through joint operation planning and the skillful assimilation of forces, capabilities, and systems to enable their employment in a single, cohesive operation rather than a set of separate operations. A synchronization matrix may be employed to visually portray critical actions that must be accomplished by multiple elements of the joint force. Coordination is facilitated through the exchange of liaisons and interoperable communications systems. These mechanisms provide the JFC with a linkage to the joint force staff and subordinate commands' activities and work to execute plans and coordinate changes required by the unfolding situation. In interagency and/or multinational environments where unity of command may not be possible, unity of effort may be achieved through effective coordination, exchange of liaisons, and interoperable communications and/or common operating systems. Constant vertical and horizontal coordination and cooperation between the CCMD and component staffs and other CCMDs are prerequisites for ensuring timely command awareness.

(7) Battle Rhythm Discipline. A command headquarters battle rhythm is its daily operations cycle for briefings, meetings, and reporting requirements. A battle rhythm is essential to support decision making, staff actions, and higher headquarters information requirements and to manage the dissemination of decisions and information in a coordinated manner. A defined battle rhythm should be based on the information requirements of the CCDR, subordinates, and senior commands. It must be designed to minimize the time the commander and key staff members spend attending meetings and listening to briefings; it must allow the staff and subordinate commanders time to plan, communicate with the commander, and direct the activities of their subordinates. The battle rhythms of the joint and component headquarters should be synchronized and take into account multiple time zones and other factors. Other planning, decision, and operating cycles or processes (intelligence, targeting, and air tasking order) influence the joint force

V-16   JP 1

Joint Command and Control

headquarters battle rhythm. Further, meetings of the necessary staff organizations must be synchronized. Consequently, key members of the joint force staff, components, and supporting agencies should participate in the development of the joint force headquarters battle rhythm. Those participants must consider the battle rhythm needs of higher, lower, lateral, and adjacent commands when developing the joint force headquarters battle rhythm.

(a) Simple, focused displays of information delivered in a disciplined way are necessary. Information displayed or discussed should be mission-related. The attention of the JFC and joint staff is pulled both from above, by requirements from seniors, and from below, by the needs of component commanders and their staffs. These requirements must also be integrated into the activities of the JFC, but must not be allowed to dominate JFC actions. Technology offers a means to reduce the time required for conducting these essential C2 events. For example, video teleconferencing and other collaborative communication tools are common methods used in many headquarters to conduct scheduled and unscheduled meetings and conferences that include a wide range of key participants.

(b) The JFC and staff must be sensitive to the battle rhythm of subordinate organizations. Component commanders also need information to function properly within their own decision cycles. The JFC should establish and require adherence to norms that increase the speed of the component commanders' decision cycles.

(8) Responsive, Dependable, and Interoperable Support Systems. ISR, space-based, and communications systems must be responsive and dependable in real time to provide the JFC with accurate, timely, relevant, and adequate information. Linking support systems that possess commonality, compatibility, and standardization to the greatest extent possible will contribute to a higher state of interoperability and thus C2 utility. Integrating the support systems of multinational and other agency partners also must be considered.

(9) Situational Awareness. The primary objective that the staff seeks to attain for the commander and for subordinate commanders is SA—a prerequisite for commanders anticipating opportunities and challenges. True situational understanding should be the basis for all decision makers. Knowledge of friendly capabilities and adversary capabilities, intentions, and likely COAs enables commanders to focus joint efforts where they best and most directly contribute to achieving objectives. Further, the JFC's SA must be broad to include the actions and intentions of multinational partners, civilian agencies, adjacent commands, higher headquarters, HN authorities, and NGOs.

(10) Mutual Trust. Decentralized execution, operating within the JFC's intent, and mission-type orders capitalize on the initiative of subordinate commanders. For these methods to work within a joint force and for the joint force to function at all, there must be a high degree of mutual trust. Trust among the commanders and staffs in a joint force expands the JFC's options and enhances flexibility, agility, and the freedom to take the initiative when conditions warrant. The JFC trusts the chain of command, leaders, and staffs to use the authority delegated to them to fulfill their responsibility for mission accomplishment; and the joint force trusts the JFC to use component capabilities

V-17

Chapter V

appropriately. Mutual trust results from honest efforts to learn about and understand the capabilities that each member brings to the joint force, demonstrated competence, and planning and training together.

b. **Decision-Making Model.** Joint operation planning occurs within the APEX system, which is the DOD-level system of joint policies, processes, procedures, and reporting structures, supported by communications and information technology that is used by the JPEC to monitor, plan, and execute mobilization, deployment, employment, sustainment, redeployment, and demobilization activities associated with joint operations. The APEX system formally integrates the planning activities of the JPEC and facilitates the JFC's seamless transition from planning to execution during times of crises. The APEX system activities span many organizational levels, but **the focus is on the interaction between SecDef and CCDRs, which ultimately helps the President and SecDef decide when, where, and how to commit US military capabilities.** The interactive and collaborative process at the national level guides the way in which planning and execution occur throughout the Armed Forces.

For further guidance on joint operation and campaign planning, refer to JP 5-0, Joint Operation Planning.

13. Organization for Joint Command and Control

Component and supporting commands' organizations and capabilities must be integrated into a joint organization that enables effective and efficient joint C2. The C2 structure is centered on the JFC's mission and CONOPS; available forces and capabilities; and joint force staff composition, capabilities, location, and facilities. The JFC should be guided in this effort by the following principles:

a. **Simplicity.** Unity of command must be maintained through an unambiguous chain of command, well-defined command relationships, and clear delineation of responsibilities and authorities. The JFC staff does not have direct authority over any subordinate commander's staffs. The component staffs work solely for the component commander.

b. **Span of Control.** The desired reach of the JFC's authority and direction over assigned or attached forces will vary depending on the mission and the JFC's ability to C2 the actions required. Span of control is based on many factors, including the number of subordinates, number of activities, range of weapon systems, force capabilities, size and complexity of the operational area, and method used to control operations (centralized or decentralized).

c. **Unit Integrity.** Component forces should remain organized as designed and in the manner accustomed through training to maximize effectiveness. However, if a JFC desires to reorganize component units, it should be done only after careful consultation and coordination with the Service component commander.

d. **Interoperability.** C2 capabilities within joint force headquarters, component commands, and supporting commands must be interoperable to facilitate control of forces.

V-18   JP 1

Joint Command and Control

The simplest and most streamlined chain of command can be thwarted by an absence of interoperability among the components' forces and systems.

14. Joint Command and Staff Process

a. **General.** The nature, scope, and tempo of military operations continually changes, requiring the commander to make new decisions and take new actions in response to these changes. This may be viewed as part of a cycle, which is repeated when the situation changes significantly. The cycle may be deliberate or rapid, depending on the time available. However, effective decision making and follow-through require that the basic process be understood by all commanders and staff officers and adapted to the prevailing situation. Although the scope and details will vary with the level and function of the command, the purpose is constant: analyze the situation and need for action; determine the COA best suited for mission accomplishment; and carry out that COA, with adjustments as necessary, while continuing to assess the unfolding situation.

b. **Estimates, Decisions, and Directives.** These processes are iterative, beginning with the initial recognition that the situation has changed (e.g., change of mission, change in the friendly or adversary situation), requiring a new decision by the commander. The staff assembles available information regarding the adversary, friendly, and environmental situations and assists the commander in analyzing the mission and devising COAs. The staff then analyzes these COAs and the commander makes a decision. This decision identifies what the command is to do and becomes the "mission" paragraph of a plan or order. An estimate process, as described in JP 5-0, *Joint Operation Planning,* may be used by commanders and staffs during the preparation of estimates and directives. Simulation and analysis capabilities can assist in correlation of friendly and adversary strengths and weaknesses, as well as in analysis of COAs.

c. **Follow-Through.** Having received and analyzed the mission, the commander determines how it will be accomplished and directs subordinate commanders to accomplish certain tasks that contribute to the common goal. Then the commander is responsible for carrying out the mission to successful conclusion, using supporting staff studies, coordination, and analysis relating to:

(1) Supervision of the conduct of operations;

(2) Changes to orders, priorities, and apportionment of support;

(3) Commitment and reconstitution of the reserve; and

(4) After mission attainment, consolidation and refit in preparation for the next task.

15. Command and Control Support

A command and control support (C2S) system, which includes interoperable supporting communications systems, is the JFC's principal tool used to collect, transport, process, share, and protect data and information. Joint C2S systems must provide quality

Chapter V

information to allow relevant and timely JFC decisions and provide feedback on the intended outcome. To facilitate the execution and processes of C2, military communications systems must furnish rapid, reliable, and secure information throughout the chain of command. All joint functions—C2, intelligence, fires, movement and maneuver, protection, sustainment, and information—depend on responsive and dependable communications systems that tie together all aspects of joint operations and allow the JFCs and their staffs to initiate, direct, monitor, question, and react. Ultimately, effective C2 depends on the right person having the right information at the right time to support decision making.

For further guidance on information quality criteria, refer to JP 6-0, Joint Communications System, *and JP 3-13,* Information Operations.

16. National Military Command System

The NMCS is the priority component of the DOD information networks designed to support the President, SecDef, and the JCS in the exercise of their responsibilities. The NMCS provides the means by which the President and SecDef can receive warning and intelligence so that accurate and timely decisions can be made, the resources of the Military Services can be applied, military missions can be assigned, and direction can be communicated to CCDRs or the commanders of other commands. Both the communication of warning and intelligence from all sources and the communication of decisions and commands to military forces require that the NMCS be a responsive, reliable, and survivable system. An enduring command structure with survivable systems is both required and fundamental to NMCS continuity of operations.

For further information, refer to JP 6-0, Joint Communications System.

17. Nuclear Command and Control System

General operational responsibility for the Nuclear Command and Control System (NCCS) lies with CJCS and is centrally directed through the Joint Staff. **The NCCS supports the Presidential nuclear C2** of the CCMDs in the areas of integrated tactical warning and attack assessment, decision making, decision dissemination, and force management and report back. To accomplish this, the NCCS comprises those critical communications system components of the DOD information networks that provide connectivity from the President and SecDef through the NMCS to the nuclear CCDRs and nuclear execution forces. It includes the emergency action message dissemination systems and those systems used for tactical warning/attack assessment, conferencing, force report back, reconnaissance, retargeting, force management, and requests for permission to use nuclear weapons. The NCCS is integral to and ensures performance of critical strategic functions of the Global Command and Control System. The Minimum Essential Emergency Communications Network provides assured communications connectivity between the President and the strategic deterrent forces in stressed environments.

V-20

JP 1

18. Defense Continuity Program

The Defense Continuity Program is an integrated program composed of DOD policies, plans, procedures, assets, and resources that ensures continuity of DOD component mission-essential functions under all circumstances, including crisis, attack, recovery, and reconstitution. It encompasses the DOD components performing continuity of operations, continuity of government, and enduring constitutional government functions to enhance readiness posture.

For further information, refer to DODD 3020.26, Department of Defense Continuity Programs.

Chapter V

Intentionally Blank

CHAPTER VI
JOINT FORCE DEVELOPMENT

"It's clear we have work to finish in the current conflicts and it should be just as clear that we have work to do in preparing for an uncertain future. Our work must result in a joint force that is responsive, decisive, versatile, interdependent, and affordable."

General Martin E. Dempsey, US Army
18th Chairman of the Joint Chiefs of Staff, 2011

SECTION A. FUNDAMENTALS OF JOINT FORCE DEVELOPMENT

1. Principles

a. Joint force development prepares individual members and units of the Armed Forces to field a joint force that integrates service capabilities in order to execute assigned missions. It includes joint doctrine, joint education, joint training, joint lessons learned, and joint concept development and assessment.

b. The *why, how,* and *what* of joint force development are:

(1) **Why** is joint force development essential? Each of the Services organizes, trains, and equips to bring unique capabilities to the fight, and the integration of these Service capabilities is the foundation of US warfighting capability. Joint force development enables the continual improvement of joint capabilities, achieving jointness at the right level. However, jointness is not automatic and it is perishable. It must be advanced through continual joint force development efforts.

(2) The overarching process for **how** jointness is maintained is the joint force development life cycle (Figure VI-1). Our joint warfighting capability is *improved* through the development of concepts validated by rigorous assessment and lessons learned from current operations. It is *sustained* through joint doctrine, education, training, and exercises. New ideas are discovered through active scouting: capitalizing and exploiting innovative opportunities and developments occurring inside and outside of the military community. The end product is joint warfighting capability.

(3) The **what** of joint force development is a trained and capable joint force.

2. Authorities

a. Joint force development involves the synchronized execution of the legislated authorities of the CJCS, the Service Chiefs, and others (such as CDRUSSOCOM). US law (Title 10, USC, Section 153) gives the CJCS authority regarding joint force development, specifically providing authority to develop doctrine for the joint employment of the Armed Forces, and to formulate policies for the joint training of the Armed Forces to include policies for the military education and training of members of the Armed Forces. In

Chapter VI

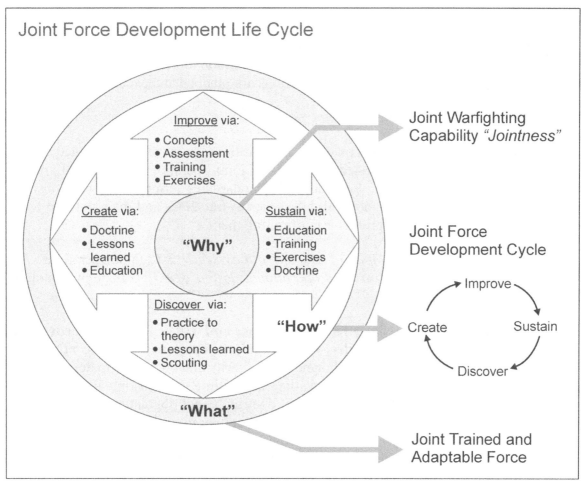

Figure VI-1. Joint Force Development Life Cycle

executing these authorities, the CJCS works in collaboration with the Service Chiefs and others (such as CDRUSSOCOM) to produce joint force capabilities.

b. These authorities derive from the Goldwater-Nichols Department of Defense Reorganization Act of 1986, which significantly increased the role of the CCDRs in many aspects of joint operations. As a result of the Goldwater-Nichols Reorganization Act, the CJCS and CCDRs have a much larger role in joint force development, especially during the planning, programming, and budgeting processes. All Services contribute their unique capabilities to the joint operation or campaign. Their integration is critical to overall joint force capability and effectiveness.

SECTION B. JOINT FORCE DEVELOPMENT PROCESS

3. Joint Force Development

Joint force development is a knowledge-based and integrated enterprise. A discussion of each of the force development subordinate processes follows.

Joint Force Development

4. Joint Doctrine

Joint doctrine provides the fundamental principles that guide the employment of US military forces in coordinated action toward a common objective. It also provides authoritative guidance from which joint operations are planned and executed.

a. **Joint Doctrine Fundamentals**

(1) Joint doctrine is based on extant capabilities (i.e., current force structures and materiel). It incorporates time-tested principles of joint operations, operational art, and elements of operational design. Joint doctrine standardizes terminology, relationships, responsibilities, and processes among all US forces to free JFCs and their staffs to focus efforts on solving the complex problems confronting them.

For more discussion of the principles of joint operations, see JP 3-0, Joint Operations. *For more discussion of operational art and operational design, see JP 5-0,* Joint Operation Planning.

(2) **Joint doctrine is authoritative guidance and will be followed except when, in the judgment of the commander, exceptional circumstances dictate otherwise.** Doctrine does not replace or alter a commander's authority and obligation to determine the proper COA under the circumstances prevailing at the time of decision; such judgments are the responsibility of the commander. Joint doctrine is not dogmatic—the focus is on how to think about operations, not what to think about operations. It is definitive enough to guide operations while versatile enough to accommodate a wide variety of situations. Joint doctrine should foster initiative, creativity, and conditions that allow commanders the freedom to adapt to varying circumstances. **The judgment of the commander based upon the situation is always paramount.**

(3) Joint doctrine applies to the Joint Staff, CCDRs, subordinate unified commanders, JTF commanders, and subordinate component commanders of these commands, the Services, and CSAs. In developing joint doctrine, existing Service, multi-Service, and multinational doctrine is considered. However, **joint doctrine takes precedence over individual Service's doctrine, which must be consistent with joint doctrine.** Joint doctrine should not include detail that is more appropriate in Service doctrine, standing operating procedures, plans, and other publications. If conflicts arise between the contents of joint doctrine and the contents of Service or multi-Service doctrine, joint doctrine takes precedence for the activities of joint forces unless CJCS has provided more current and specific guidance.

(4) **Joint doctrine is not policy.** Policy and doctrine are closely related, but they fundamentally fill separate requirements. Policy can direct, assign tasks, prescribe desired capabilities, and provide guidance for ensuring the Armed Forces of the United States are prepared to perform their assigned roles; implicitly policy can create new roles and a requirement for new capabilities. Most often, policy drives doctrine; however, on occasion, an extant capability will require policy to be created. As doctrine reflects extant

VI-3

Chapter VI

capabilities, policy must first be implemented and/or new capabilities fielded before they can be written into doctrine.

(5) When the Armed Forces of the United States participate in multinational operations, US commanders should follow multinational doctrine and procedures that were ratified by the US. For multinational doctrine and procedures not ratified by the US, commanders should evaluate and follow the multinational command's doctrine and procedures where applicable and consistent with US law, policy, and doctrine.

(6) Joint doctrine is developed under the aegis of the CJCS in coordination and consultation with the Services, CCMDs, and CSAs. The Joint Staff leads the joint doctrine development community and is responsible for all aspects of the joint doctrine process, to include promulgation.

For further guidance on the development of joint doctrine, refer to CJCSI 5120.02C, Joint Doctrine Development System.

b. **Purpose of Joint Doctrine**

Joint doctrine is written for those who:

(1) Provide strategic direction to joint forces (the CJCS and CCDRs).

(2) Employ joint forces (CCDRs, subordinate unified commanders, or JTF commanders).

(3) Support or are supported by joint forces (CCMDs, subunified commands, JTFs, component commands, the Services, and CSAs).

(4) Prepare forces for employment by CCDRs, subordinate unified commanders, and JTF commanders.

(5) Train and educate those who will conduct joint operations.

5. **Joint Education**

a. Education is a key aspect of the joint force development process.

(1) Professional military education (PME) conveys the broad body of knowledge and develops the cognitive skills essential to the military professional's expertise in the art and science of war. Additionally, affective or attitudinal learning is paired with education to better inculcate the values of joint service as discussed in Chapter I, "Theory and Foundations."

(2) Service delivery of PME, taught in a joint context, instills basic Service core competency within topics associated with joint matters. Joint education is the aspect of PME that focuses on imparting joint knowledge and attitudes.

VI-4

JP 1

(3) Joint education can be broadly parsed into three categories:

(a) **Joint Professional Military Education (JPME).** As codified in US law (Title 10, USC, Section 2151) JPME "consists of the rigorous and thorough instruction and examination of officers of the armed forces in an environment designed to promote a theoretical and practical understanding of joint matters." JPME fulfills the educational requirements of the 1986 Goldwater-Nichols Act. JPME is delivered in both service and purely joint (e.g., the National Defense University) venues. JPME positions officers to recognize and operate in tactical, operational, and strategic levels of national security. JPME curricula—based upon approved joint doctrine and concepts—address what every officer should know regarding joint matters at appropriate levels from pre-commissioning through general/flag officer.

For further guidance on JPME, refer to CJCSI 1800.01D, Officer Professional Military Education Policy (OPMEP).

(b) **Enlisted Joint Professional Military Education (EJPME).** Although not mandated in US law, the joint development of enlisted personnel requires some degree of joint education. EJPME is largely delivered in Service PME venues along two separate tracks. The first track addresses what every enlisted member of the Armed Forces should know regarding joint subjects at appropriate levels of service; the second track is specific educational preparation for enlisted members for joint duty. Like JPME, EJPME curricula are based upon approved joint doctrine and concepts.

For further guidance on EJPME, refer to CJCSI 1805.01A, Enlisted Professional Military Education Policy.

(c) **Other Joint Education.** JPME and EJPME provide members of the Armed Forces with common education on joint matters. JPME and EJPME curricula provide "some joint content for all"—what all members of the Armed Forces should know. Other joint education provides specialist education in specific joint functional areas. Other joint education (sometimes called "small j" joint education to distinguish it from JPME/EJPME) is topic-specific education of joint material for both officers and enlisted personnel. The content of "other joint education," albeit significant to joint officer development, does not otherwise meet standards to be accredited as JPME. This type of education is sponsored by topic-specific relevant Services and organizations (i.e., joint logistics education by the joint logistics community). Other joint education conveys a specific body of knowledge relevant to the specific field and is also based on approved joint doctrine and concepts.

b. **Influence of Joint Education**

(1) Joint education is based on joint doctrine and should reflect the deliberate, iterative, and continuous nature of joint force development. Joint curricula should include approved joint concepts and the most recent observed lessons from across the joint force.

(2) Joint education is closely related to individual joint training. Policy recognizes that education and training are not mutually exclusive. Virtually all military

Chapter VI

schools and professional development programs include elements of both education and training in their academic curriculum. Effective joint learning relies on close coordination of training and education. As individuals mature and develop within their military specialties, they acquire the knowledge, skills, and abilities required for positions of increased responsibilities.

6. Joint Training

Joint training prepares individuals, joint forces, or joint staffs to respond to strategic, operational, or tactical requirements considered necessary by the CCDRs to execute their assigned or anticipated missions. Joint training encompasses both individual and collective training of joint staffs, units, and the Service components of joint forces.

a. Joint Training Fundamentals

(1) **Types of Joint Training.** Joint training must be accomplished by effectively matching the training requirements and training audiences with appropriate training methods and modes within available resources. These audiences include:

(a) **Individual Joint Training.** Training that prepares individual members and commanders to perform duties in joint organizations (e.g., specific staff positions or functions) or to operate uniquely joint systems. Individuals should be proficient in requisite knowledge, skills, and ability to apply joint doctrine and procedures necessary to function as staff members.

(b) **Staff Joint Training.** Training that prepares joint staffs or joint staff elements to respond to strategic and operational taskings deemed necessary by CCDRs and subordinate JFCs to execute their assigned missions.

(c) **Collective Joint Training.** Instruction and applied exercises that prepare joint organizational teams to integrate and synchronize owned and provided capabilities to execute assigned missions. Collective exercise programs include the President's National Exercise Program (NEP), the Chairman's Exercise Program (CEP), and the Joint Exercise Program (JEP).

1. The NEP consists of annual, operations-based exercises, either a functional command post exercise or a full-scale exercise, involving department and agency principals and Presidential participation. These national-level exercises address USG strategic and policy-level objectives and challenge the national response system. DOD participates in the NEP through the CEP.

2. The CEP is the only dedicated means for the CJCS, through the Joint Staff, to coordinate interagency and CCMD participation in strategic national-level joint exercises.

3. The JEP is a principal means for CCDRs to maintain trained and ready forces, exercise their contingency plans, support their TCP, and achieve joint and multinational (combined) training. CCDR-sponsored JEP events train to mission

VI-6   JP 1

Joint Force Development

capability requirements described in the command joint mission-essential task list (JMETL) as well as theater security cooperation requirements as directed in TCPs.

(2) **Tenets of Joint Training.** JFCs must integrate and synchronize the actions of their forces to achieve strategic and operational objectives. Success depends on well-integrated command headquarters, supporting organizations, and forces that operate as a team. The tenets of joint training are intended to guide commanders and agency directors in developing their joint training plans: use joint doctrine; commanders/agency directors are the primary trainers; mission focus; train the way you intend to operate; centralize planning, decentralize execution; and link training and readiness assessments.

(3) **End State.** The desired end state of joint training is a training and exercise strategy aligned with the NMS that results in joint force readiness.

(4) **Joint Training System (JTS).** The JTS is a four-phased methodology that aligns training strategy with assigned missions to produce trained and ready individuals, units, and staffs.

(a) The first phase (the Requirements Phase) identifies the required capabilities identified during mission analyses by commanders or agency directors and their staffs, and are based on assigned mission responsibilities, commander's intent, and joint doctrine and documented in the command/agency JMETL or agency mission-essential task list (AMETL).

(b) The second phase (the Plans Phase) is an analysis of current capabilities against required capabilities (JMETL/AMETL). Joint training requirements are derived by analyzing gaps between mission capability requirements and current capability proficiency. Training methods, modes, and media are determined based on the desired level of performance, which determines the type of training events required.

(c) The third phase (the Execution Phase) refines and finalizes, executes, and evaluates training events scheduled during the plans phase. Following execution, command/agency trainers collect the task performance observations for each training objective, conduct analysis, and make a formal recommendation as training proficiency evaluations on whether the training audience achieved the training objective.

(d) The fourth phase (the Assessments Phase) completes the joint training cycle and begins the next cycle as an input to future training plans. It focuses on the organization's capability to accomplish its assigned missions. It may also impact near-term training if critical shortcomings or deficiencies in a command's proficiency, or in overall joint procedures, are identified.

(5) Although the process is deliberate in concept, it is flexible in execution. The JTS is a systematic approach to assist commanders in ensuring readiness levels required in their command are met through effective joint training. This approach assists in identifying the functional responsibilities of assigned individuals and organizations in the form of tasks, conditions, and standards; identifying events and resources to accomplish required

Chapter VI

training; conducting and evaluating training audience performance; and assessing their ability to perform assigned mission tasks in the training environment.

(6) The JTS supports DOD implementation of the joint learning continuum by providing the construct to plan and implement a comprehensive organizational program that may include elements of training (individual, staff, and collective), education, self-development, and experience to achieve mission capability. This process establishes the linkage between the NMS and CCMD missions, operational plans, and tasks inherent in those plans and joint training. The final product is a CCDR JMETL and the CSA AMETL that reflects and consolidates the mission capability requirements of the CCMD into a single list of tasks, conditions, standards, responsible individuals, and organizations.

(7) The JTS represents a series of logical and repeatable processes that are intended to continuously improve joint readiness. Used correctly, the system helps CCDRs, subordinate joint force, functional, or Service commanders, other senior commanders, and CSA directors to train more efficiently and identify areas for improvement. Effectively using the processes within the JTS better enables commanders to assess the level of training readiness in their commands and then make informed judgments on their ability to perform assigned missions under unified CCMD.

b. **Influence of Joint Training**

(1) The Universal Joint Task List (UJTL) is a doctrine-based construct detailing a universe of military tasks linked to specific conditions and standards. The UJTL provides a standardized tool to support the planning, execution, evaluation, and assessment of joint training. The UJTL offers a library of mission tasks for the development of JMETLs/AMETLs and readiness reporting as well as defining capability requirements for readiness reporting, systems acquisition, and contingency operations planning.

(2) The ultimate objective of the joint learning continuum is the provision of mission-ready individuals and collective entities. Broadly, the missions to be accomplished are articulated in the war plans of the CCDRs.

For further guidance on joint training, refer to CJCSI 3500.01G, Joint Training Policy and Guidance for the Armed Forces of the United States.

7. **Lessons Learned**

a. The joint lessons learned component of joint force development entails collecting observations, analyzing them, and taking the necessary steps to turn them into "learned lessons"—changes in behavior that improve the mission ready capabilities of the joint force. The conduct of joint operations provides the truest test of joint doctrine, joint education, and joint training. Accordingly, it is crucially important to observe keenly the conduct of joint operations, as well as the execution of each part of the joint force development process, in order to continuously identify and assess the strengths and weaknesses of joint doctrine, joint education, and joint training as well as strategy, policy, materiel, and supporting military systems. Properly assessed, these positive and negative observations help senior leaders identify and fix problems, reinforce success, and inside the joint force development perspective, adjust the azimuth and interaction of the various lines of effort.

b. **The Joint Lessons Learned Program (JLLP)** is focused on improving joint preparedness and performance. Its primary objective is to enhance our abilities to conduct joint operations by contributing to improvements in doctrine, organization, training, materiel, leadership and education, personnel, facilities, and policy. It enhances joint force capabilities by enabling learning and collaboration from joint activities, including engagement, planning, training, exercises, experiments, operations, real-world events, and other activities involving the Armed Forces of the United States. The JLLP improves performance through discovery, evaluation, validation, and integration with learning and resolution processes, to ensure lessons are learned and integrated across DOD. Learned lessons are the building blocks that feed the revisions and updates to doctrine, education, and training. They identify gaps and common issues, and lead to the capture and implementation of best practices and experiences of DOD forces. Ultimately, this leads to the development of new capabilities and improvements in the readiness of our forces. The JLLP is designed to support the USG whole-of-government effort by sharing and collaborating learned lessons information with other USG organizations and multinational partners.

For further guidance on the JLLP, refer to CJCSI 3150.25E, Joint Lessons Learned Program.

8. Joint Concepts and Assessment

a. Joint concepts examine military problems and propose solutions describing how the joint force, using military art and science, may operate to achieve strategic goals within the context of the anticipated future security environment. Joint concepts lead to military capabilities, both non-materiel and materiel, that significantly improve the ability of the joint force to overcome future challenges.

b. **Joint Concept Fundamentals**

(1) **Joint concepts provide solutions** to compelling, real-world challenges both current and envisioned for which existing doctrinal approaches and joint capabilities are deemed inadequate. The absence of doctrine may indicate that the joint force has encountered a situation with which there has been no previous experience. As battlefield conditions, technology, and opposing force capabilities evolve, concept development provides a means to address these challenges. Concepts proceed from an understanding of existing doctrine or knowledge of existing capabilities. They must propose a clear alternative to existing doctrine or augmentation of existing capabilities and demonstrate evidence of significant operational value relative to the challenges under consideration.

(2) **Joint concepts are idea-focused** and are not constrained by existing policies, treaties, laws, or technology. This permits the development of concepts that anticipate conditions as they may exist in the future. In this way it is possible to start with an idea—a visualization of how forces could successfully operate against specific challenges and

Chapter VI

across the joint functions—and proceed to describe new employment methodologies for existing capabilities as well as new capability requirements.

(3) **Joint concept development and assessment** is focused on mitigating DOD's highest joint force development needs. Using joint force development priorities as determined by the CJCS, CCMDs and Services identify and nominate proposals to solve current or anticipated capability gaps annually via the Comprehensive Joint Assessment. Once nominated these proposals are consolidated, reviewed, and assigned a proposed priority. A proposed list of projects and associated priority are reviewed by the Joint Capabilities Board and forwarded to the JROC for approval. Upon approval by the JROC this list becomes the joint concept development and experimentation annual program of work.

(4) **Joint concepts are developed collaboratively** to ensure that a wide range of ideas and perspectives are considered for developing a deep understanding of the problem and generating options for solving that problem. Military leaders and other subject matter experts from a variety of backgrounds collectively assess the challenge and identify potential solutions. The process employs a systematic methodology for incorporating real-world observations, concerns, or issues as well as consideration of a wide range of innovative ideas for more effective future operations.

(5) **Joint concepts are rigorously evaluated.** Joint concepts are objectively evaluated to ensure that proposed solutions will enable the joint force to successfully overcome the operational challenges the concept was intended to address. The validation process involves testing the ideas during development; first in assessments under controlled environments and then in operational environments, such as exercises.

c. **Joint Concept Assessment.** A joint assessment is an analytical activity based on unbiased trials conducted under controlled conditions within a representative environment, to validate a concept, hypothesis, discover something new, or establish knowledge. Results of an assessment are reproducible and provide defensible analytic evidence for joint force development decisions. Assessments and wargames provide the means to determine the efficacy of proposed capability to challenges, problems, and issues facing CCMDs, Services, agencies, and MNFs.

(1) Joint assessment uses a variety of methodologies to develop and evaluate capabilities to mitigate the joint force's most pressing challenges. These various approaches enable joint force development by increasing understanding, developing knowledge, or improving capabilities. These approaches range from discovery events, workshops, and seminars to hypothesis testing and demonstrations. Regardless of the approach the goal is to produce operationally relevant, credible, defensible, and sustainable joint capabilities.

(2) Joint assessment is planned and executed in a collaborative five phase cycle. This cycle uses a find, plan, experiment, finish, and exploit process model, incorporating continuous project and program assessment.

VI-10   JP 1

Joint Force Development

d. **Influence of Joint Concepts and Assessment**

(1) Joint concepts and validated results from joint assessments are deliberately transitioned to non-materiel capability development systems to become sustainable joint capabilities. Once validated and approved, concepts and solutions are actively transitioned via the force development process in order to institutionalize the solutions. Succeeding in this requires the establishment of three types of organizational sponsorship to deliver sustainable joint capability to the joint warfighter. The operational sponsor is the submitting organization ("has the problem"). The technical sponsor is the organization that has the expertise in designing and executing the experiment or expertise with the proposed capability. The implementation sponsor is the organization that has the authorities and responsibilities required for implementation of the developed capability, once validated and approved, into the joint force. These roles and responsibilities are identified in a project experimentation support agreement during the planning phase and serve as the baseline for formal periodic project reviews and ensure that assessment remains relevant to DOD.

(2) Joint assessment does not conclude the joint force development life cycle; it informs it. Being that the joint force development life cycle is iterative, constant, and inclusive, results of assessments provide validated solutions that provide change recommendations for relevant joint doctrine, education, training, and exercises to sustain the joint warfighter's capability. Additionally, as new capabilities are discovered through the collection and exploration of lessons learned, joint warfighting capability is created with the codification of these best practices into joint doctrine, the dissemination of lessons learned across the Services, and comprehensive training and education programs that produce our future strategists and leaders. Accordingly, joint force development enables a joint trained and adaptable force prepared to function across the ROMO.

For further guidance on joint concepts and experimentation, refer to CJCSI 3010.02C, Joint Concept Development and Experimentation.

Chapter VI

Intentionally Blank

APPENDIX A
ESTABLISHING DIRECTIVE (SUPPORT RELATIONSHIP) CONSIDERATIONS

1. General

The following information is provided to assist CCDRs, subordinate JFCs, and other commanders with the authority to designate a support relationship between subordinate commanders and with considerations in developing an establishing directive to clarify that support relationship.

2. Establishing Directive

a. An establishing directive is essential to ensure unity of command. Normally, the designated commander will develop a draft establishing directive during the planning phase to provide the specifics of the support relationship. The commander will submit the draft establishing directive to the establishing authority for consideration. The establishing directive is normally issued to specify the purpose of the support relationship, the effects desired, the objectives, and the scope of the action to be taken. It may also include but is not necessarily limited to the following:

(1) Time, place, level, and duration of the supporting effort.

(2) Relative priority of the supporting effort.

(3) Authority, if any, of the supporting commanders to modify the supporting effort in the event of exceptional opportunity or an emergency.

(4) Degree of authority granted to the supported commander over the supporting effort.

(5) Establishment of air, maritime, ground, and cyberspace maneuver control measures.

(6) Development of joint TACAIR strike requests and air support requests.

(7) Development of target nominations, establishment of fire support coordination measures, integration of air defense, and the role of coordination centers.

(8) Development of the current enemy situation, joint intelligence preparation of the operational environment to guide the joint operation planning process, intelligence collection plan, and ISR strategy.

(9) Nonorganic logistic support.

(10) Force protection responsibilities.

Appendix A

b. Unless otherwise stated in the establishing directive, the supported and supporting commanders will identify the events and conditions for any shifts of the support relationship throughout the operation during the planning phase and forward them to the establishing authority for approval. The establishing authority will resolve any differences among the commanders.

3. Supported Commander

A supported commander may be designated for the entire operation, a particular phase or stage of the operation, a particular function, or a combination of phases, stages, events, and functions. Unless limited by the establishing directive, **the supported commander has the authority to exercise general direction of the supporting effort.** General direction includes the designation and prioritization of targets or objectives, timing and duration of the supporting action, and other instructions necessary for coordination and efficiency. The establishing authority is responsible for ensuring that the supported and supporting commanders understand the degree of authority that the supported commander is granted.

a. If not specified, the establishing authority (the common superior commander) will determine who has primary responsibility for the essential tasks during the mission analysis in the planning process.

b. **In an operation of relatively short duration, normally the establishing authority will choose one supported commander for the entire operation.**

4. Supporting Commander

The supporting commander determines the forces, tactics, methods, procedures, and communications to be employed in providing this support. The supporting commander will advise and coordinate with the supported commander on matters concerning the employment and limitations (e.g., logistics) of such support, assist in planning for the integration of such support into the supported commander's effort as a whole, and ensure that support requirements are appropriately communicated throughout the supporting commander's organization. The supporting commander has the responsibility to ascertain the needs of the supported force and take full action to fulfill them within existing capabilities, consistent with priorities and requirements of other assigned tasks. When the supporting commander cannot fulfill the needs of the supported commander, the establishing authority will be notified by either the supported or supporting commanders. The establishing authority is responsible for determining a solution.

APPENDIX B
THE PROFESSION OF ARMS

1. General

a. A professional is a person of **both** character **and** competence. As military professionals charged with the defense of the Nation, joint leaders must be experts in the conduct of war. They must be moral individuals both of action and of intellect, skilled at getting things done, while at the same time conversant in the military art.

b. Every joint leader is expected to be a student of the art and science of war. Officers especially are expected to have a solid foundation in military theory and philosophy, and knowledge of military history and the timeless lessons to be gained from it. Leaders must have a strong sense of the great responsibility of their office; the resources they will expend in war include their fellow citizens.

c. Strong character and competence represent the essence of the US joint military force and its leaders. Both are the products of lifelong learning and are embedded in JPME.

2. Character and Competence

a. **Character** refers to the aggregate of features and traits that form the individual nature of a person. In the context of the profession of arms, it entails moral and ethical adherence to our values. Character is at the heart of the relationship of the profession with the American people, and to each other.

b. **Competence** is central to the profession of arms. Competent performance includes both the technical competence to perform the relevant task to standard as well as the ability to integrate that skill with others. Those who will lead joint operations must develop skill in integrating forces into smoothly functioning joint teams.

c. **Values of Joint Service.** US military service is based on values that US military experience has proven to be vital for operational success. These values adhere to the most idealistic societal norms, are common to all the Services, and represent the essence of military professionalism. Integrity, competence, physical courage, moral courage, and teamwork all have special impact on the conduct of joint operations.

3. Values

a. US military service is based on values that US military experience has proven to be vital for operational success. These values adhere to the most idealistic societal norms, are common to all the Services, and represent the essence of military professionalism. **Duty, honor, courage, integrity, and selfless service** are the calling cards of the profession of arms.

(1) **Duty** is our foremost value. It binds us together and conveys our moral commitment or obligation as defenders of the Constitution and servants of the Nation. As

Appendix B

members of the profession of arms, we fulfill our duty without consideration of self-interest, sacrificing our lives if needed. From duty comes responsibility.

(2) **Honor** is the code of behavior that defines the ethical fulfillment of our duties. It is that quality that guides us to exemplify the ultimate in ethical and moral behavior; never to lie, cheat, or steal; to abide by an uncompromising code of integrity; to respect human dignity; to have respect and concern for each other. The quality of maturity; dedication, trust, and dependability that commits members of the profession of arms to act responsibly; to be accountable for actions; to fulfill obligations; and to hold others accountable for their actions.

(3) **Courage.** The United States of America is blessed with Soldiers, Marines, Sailors, Airmen, and Coast Guardsmen whose courage knows no boundaries. Even in warfare characterized by advanced technology, individual fighting spirit and courage remain essential. Courage has both physical and moral aspects and encompasses both bravery and fortitude.

(a) **Physical courage** has throughout history defined warriors. It is the ability to confront physical pain, hardship, death, or threat of death. Physical courage in a leader is most often expressed in a willingness to act, even alone if necessary, in situations of danger and uncertainty.

(b) **Moral courage** is the ability to act rightly in the face of popular opposition, or discouragement. This includes the willingness to stand up for what one believes to be right even if that stand is unpopular or contrary to conventional wisdom. This involves risk taking, tenacity, and accountability.

(4) **Integrity** is the quality of being honest and having strong moral principles. Integrity is the bedrock of our character and the cornerstone for building trust. Trust is an essential trait among Service members—trust by seniors in the abilities of their subordinates and by juniors in the competence and support of their seniors. American Service members must be able to rely on each other, regardless of the challenge at hand; they must individually and collectively say what they mean, and do what they say.

(5) **Selfless service** epitomizes the quality of putting our Nation, our military mission(s), and others before ourselves. Members of the profession of arms do not serve to pursue fame, position, or money. They give of themselves for the greater good. Selfless service is the enabler of teamwork, the cooperative effort by the members of a group to achieve common goals.

b. Members of the Armed Forces of the United States must internalize and embody these values of the profession of arms; their adherence to these values helps promulgate an attitude about joint warfighting, producing a synergy that multiplies the effects of their individual actions.

4. Teamwork

a. The Armed Forces of the United States—every military organization to the lowest level—are a team. Deterring adversaries and winning the Nation's wars are the team's common goals.

(1) **Trust and confidence** are central to unity of effort. A highly effective team is based on the team members having trust and confidence in each other.

(2) Successful teamwork requires **delegation of authority** commensurate with responsibility. This is a necessary part of building and maintaining the trust based on competence that characterizes the successful team. Delegation unleashes the best efforts and greatest initiative among all members of military teams.

(3) Successful teamwork also requires **cooperation.** While this aspect of teamwork can be at tension with competition, and both are central human characteristics, the nature of modern warfare puts a premium on cooperation within the team in order to prevail.

For additional details on the military profession and values, see the Chairman's white paper, America's Military—A Profession of Arms.

Appendix B

Intentionally Blank

APPENDIX C
REFERENCES

The development of JP 1 is based upon the following primary references.

1. United States Laws

a. The National Security Act of 1947, as amended.

b. Titles 10 and 32, USC, as amended.

c. Title 14, USC, Sections 1, 2, and 141.

d. The Goldwater-Nichols Department of Defense Reorganization Act of 1986.

e. Posse Comitatus Act (Title 18, USC, Section 1385).

2. Strategic Guidance and Policy

a. *Guidance for Employment of the Force.*

b. *Joint Strategic Capabilities Plan.*

c. *The National Security Strategy of the United States.*

d. *The National Defense Strategy of the United States.*

e. *The National Military Strategy of the United States.*

f. *National Strategy for Homeland Security.*

g. *National Strategy for Combating Terrorism.*

h. *National Strategy to Combat Weapons of Mass Destruction.*

i. *Unified Command Plan.*

j. *National Response Framework.*

k. *DOD Strategy for Homeland Defense and Civil Support.*

3. Department of Defense Publications

a. DODD 1200.17, *Managing the Reserve Components as an Operational Force.*

b. DODD 2000.12, *DOD Antiterrorism (AT) Program.*

c. DODD 3000.06, *Combat Support Agencies.*

Appendix C

d. DODD 3000.07, *Irregular Warfare.*

e. DODD 3020.26, *Department of Defense Continuity Programs.*

f. DODD 3020.40, *DOD Policy and Responsibilities for Critical Infrastructure.*

g. DODD 3025.12, *Military Assistance for Civil Disturbances.*

h. DODD 3025.18, *Defense Support of Civil Authorities (DSCA).*

i. DODD 5100.01, *Functions of the Department of Defense and its Major Components.*

j. DODD 5100.03, *Support of the Headquarters of Combatant and Subordinate Unified Commands.*

k. DODD 5100.20, *National Security Agency/Central Security Service (NSA/CSS).*

l. DODD 5101.1, *DOD Executive Agent.*

m. DODD 5105.19, *Defense Information Systems Agency (DISA).*

n. DODD 5105.21, *Defense Intelligence Agency (DIA).*

o. DODD 5105.22, *Defense Logistics Agency (DLA).*

p. DODD 5105.77, *National Guard Bureau (NGB).*

q. DODD 5106.4, *Combatant Command Inspectors General.*

r. DODI 1215.06, *Uniform Reserve, Training, and Retirement Categories.*

s. DODI 3000.05, *Stability Operations.*

t. DODI 4000.19, *Interservice and Intragovernmental Support.*

4. Chairman of the Joint Chiefs of Staff Publications

a. CJCSI 1800.01D, *Officer Professional Military Education Policy (OPMEP).*

b. CJCSI 1805.01A, *Enlisted Professional Military Education Policy.*

c. CJCSI 3100.01B, *Joint Strategic Planning System.*

d. CJCSI 3405.01, *Chairman's Total Force Fitness Framework.*

e. CJCSI 5120.02B, *Joint Doctrine Development System.*

f. CJCSI 5715.01B, *Joint Staff Participation in Interagency Affairs.*

References

g. CJCSM 3122.01A, *Joint Operation Planning and Execution System (JOPES), Volume I (Planning Policies and Procedures)*.

h. CJCSM 3500.03C, *Joint Training Manual for the Armed Forces of the United States*.

i. JP 1-0, *Joint Personnel Support*.

j. JP 1-02, *Department of Defense Dictionary of Military and Associated Terms*.

k. JP 1-04, *Legal Support to Military Operations*.

l. JP 1-05, *Religious Affairs in Joint Operations*.

m. JP 2-0, *Joint Intelligence*.

n. JP 3-0, *Joint Operations*.

o. JP 3-05, *Special Operations*.

p. JP 3-07.2, *Antiterrorism*.

q. JP 3-08, *Interorganizational Coordination During Joint Operations*.

r. JP 3-13, *Information Operations*.

s. JP 3-13.2, *Military Information Support Operations*.

t. JP 3-16, *Multinational Operations*.

u. JP 3-27, *Homeland Defense*.

v. JP 3-28, *Civil Support*.

w. JP 3-30, *Command and Control for Joint Air Operations*.

x. JP 3-31, *Command and Control for Joint Land Operations*.

y. JP 3-32, *Command and Control for Joint Maritime Operations*.

z. JP 3-33, *Joint Task Force Headquarters*.

aa. JP 3-57, *Civil-Military Operations*.

bb. JP 3-61, *Public Affairs*.

cc. JP 4-0, *Joint Logistics*.

dd. JP 5-0, *Joint Operation Planning*.

Appendix C

ee. JP 6-0, *Joint Communications System.*

APPENDIX D
ADMINISTRATIVE INSTRUCTIONS

1. User Comments

Users in the field are highly encouraged to submit comments on this publication to: Joint Staff J-7, Deputy Director, Joint Education and Doctrine, ATTN: Joint Doctrine Analysis Division, 116 Lake View Parkway, Suffolk, VA 23435-2697. These comments should address content (accuracy, usefulness, consistency, and organization), writing, and appearance.

2. Authorship

The lead agent and Joint Staff doctrine sponsor for this publication is the Directorate for Joint Force Development (J-7).

3. Supersession

This publication supersedes JP 1, *Doctrine for the Armed Forces of the United States*, 02 May 2007, incorporating Change 1, 20 March 2009.

4. Change Recommendations

a. Recommendations for urgent changes to this publication should be submitted:

TO: JOINT STAFF WASHINGTON DC//J7-JE&D//

b. Routine changes should be submitted electronically to the Deputy Director, Joint Education and Doctrine, ATTN: Joint Doctrine Analysis Division, 116 Lake View Parkway, Suffolk, VA 23435-2697, and info the lead agent and the Director for Joint Force Development, J-7/JE&D.

c. When a Joint Staff directorate submits a proposal to the CJCS that would change source document information reflected in this publication, that directorate will include a proposed change to this publication as an enclosure to its proposal. The Services and other organizations are requested to notify the Joint Staff J-7 when changes to source documents reflected in this publication are initiated.

5. Distribution of Publications

Local reproduction is authorized and access to unclassified publications is unrestricted. However, access to and reproduction authorization for classified JPs must be in accordance with DOD Manual 5200.01, Volume 1, *DOD Information Security Program: Overview, Classification, and Declassification,* and DOD Manual 5200.01, Volume 3, *DOD Information Security Program: Protection of Classified Information.*

Appendix D

6. Distribution of Electronic Publications

a. Joint Staff J-7 will not print copies of JPs for distribution. Electronic versions are available on JDEIS at https://jdeis.js.mil (NIPRNET) and http://jdeis.js.smil.mil (SIPRNET) and on the JEL at http://www.dtic.mil/doctrine (NIPRNET).

b. Only approved JPs and joint test publications are releasable outside the CCMDs, Services, and Joint Staff. Release of any classified JP to foreign governments or foreign nationals must be requested through the local embassy (Defense Attaché Office) to DIA, Defense Foreign Liaison/IE-3, 200 MacDill Blvd., Joint Base Anacostia-Bolling, Washington, DC 20340-5100.

c. JEL CD-ROM. Upon request of a joint doctrine development community member, the Joint Staff J-7 will produce and deliver one CD-ROM with current JPs. This JEL CD-ROM will be updated not less than semi-annually and when received can be locally reproduced for use within the CCMDs, Services, and combat support agencies.

GLOSSARY
PART I—ABBREVIATIONS AND ACRONYMS

AC	Active Component
ADCON	administrative control
AMETL	agency mission-essential task list
AOR	area of responsibility
APEX	Adaptive Planning and Execution
C2	command and control
C2S	command and control support
CA	civil affairs
CBP	capabilities-based planning
CBRN CM	chemical, biological, radiological, and nuclear consequence management
CCDR	combatant commander
CCMD	combatant command
CCS	commander's communication synchronization
CDRUSNORTHCOM	Commander, United States Northern Command
CDRUSSOCOM	Commander, United States Special Operations Command
CEP	Chairman's Exercise Program
CJCS	Chairman of the Joint Chiefs of Staff
CJCSI	Chairman of the Joint Chiefs of Staff instruction
CJCSM	Chairman of the Joint Chiefs of Staff manual
CMOC	civil-military operations center
CNGB	Chief, National Guard Bureau
COA	course of action
COCOM	combatant command (command authority)
CONOPS	concept of operations
COS	chief of staff
CSA	combat support agency
DC	Deputies Committee
DCMA	Defense Contract Management Agency
DHS	Department of Homeland Security
DIA	Defense Intelligence Agency
DIRLAUTH	direct liaison authorized
DISA	Defense Information Systems Agency
DLA	Defense Logistics Agency
DOD	Department of Defense
DODD	Department of Defense directive
DODI	Department of Defense instruction
DON	Department of the Navy
DOS	Department of State
DSCA	defense support of civil authorities
DTRA	Defense Threat Reduction Agency

Glossary

EA	executive agent
EJPME	enlisted joint professional military education
FCC	functional combatant commander
GCC	geographic combatant commander
GCP	global campaign plan
GEF	Guidance for Employment of the Force
GFMIG	Global Force Management Implementation Guidance
HD	homeland defense
HN	host nation
HS	homeland security
HSC	Homeland Security Council
IGO	intergovernmental organization
IO	information operations
IPC	interagency policy committee
ISR	intelligence, surveillance, and reconnaissance
IW	irregular warfare
JCA	joint capability area
JCS	Joint Chiefs of Staff
JDOMS	Joint Director of Military Support
JEP	Joint Exercise Program
JFACC	joint force air component commander
JFC	joint force commander
JIACG	joint interagency coordination group
JLLP	Joint Lessons Learned Program
JMETL	joint mission-essential task list
JP	joint publication
JPEC	joint planning and execution community
JPME	joint professional military education
JROC	Joint Requirements Oversight Council
JSCP	Joint Strategic Capabilities Plan
JSPS	Joint Strategic Planning System
JTF	joint task force
JTS	Joint Training System
MAGTF	Marine air-ground task force
MCM	Manual for Courts-Martial
MISO	military information support operations
MNF	multinational force
MNFC	multinational force commander
MOA	memorandum of agreement
MWR	morale, welfare, and recreation

GL-2   JP 1

NATO	North Atlantic Treaty Organization
NCCS	Nuclear Command and Control System
NDS	national defense strategy
NEP	National Exercise Program
NGA	National Geospatial-Intelligence Agency
NGB	National Guard Bureau
NGO	nongovernmental organization
NMCS	National Military Command System
NMS	national military strategy
NORAD	North American Aerospace Defense Command
NSA	National Security Agency
NSC	National Security Council
NSS	national security strategy
OPCON	operational control
OSD	Office of the Secretary of Defense
PA	public affairs
PC	Principals Committee
PME	professional military education
RC	Reserve Component
RCM	Rules for Courts-Martial
ROE	rules of engagement
ROMO	range of military operations
RUF	rules for the use of force
SA	situational awareness
SecDef	Secretary of Defense
SO	special operations
SOF	special operations forces
TACAIR	tactical air
TACON	tactical control
TCP	theater campaign plan
TFF	total force fitness
TRO	training and readiness oversight
UCMJ	Uniform Code of Military Justice
UCP	Unified Command Plan
UJTL	Universal Joint Task List
UN	United Nations
USC	United States Code
USCG	United States Coast Guard
USELEMNORAD	United States Element, North American Aerospace Defense Command

Glossary

USEUCOM	United States European Command
USG	United States Government
USNORTHCOM	United States Northern Command
USPACOM	United States Pacific Command
USSOCOM	United States Special Operations Command
VCJCS	Vice Chairman of the Joint Chiefs of Staff
WMD	weapons of mass destruction

PART II—TERMS AND DEFINITIONS

accountability. The obligation imposed by law or lawful order or regulation on an officer or other person for keeping accurate record of property, documents, or funds. (Approved for incorporation into JP 1-02 with JP 1 as the source JP.)

administrative control. Direction or exercise of authority over subordinate or other organizations in respect to administration and support. Also called **ADCON.** (Approved for incorporation into JP 1-02.)

area of responsibility. The geographical area associated with a combatant command within which a geographic combatant commander has authority to plan and conduct operations. Also called **AOR.** (JP 1-02. SOURCE: JP 1)

Armed Forces of the United States. A term used to denote collectively all components of the Army, Marine Corps, Navy, Air Force, and Coast Guard (when mobilized under Title 10, United States Code, to augment the Navy). (Approved for incorporation into JP 1-02.)

casual. None. (Approved for removal from JP 1-02.)

chain of command. The succession of commanding officers from a superior to a subordinate through which command is exercised. Also called **command channel.** (Approved for incorporation into JP 1-02.)

combatant command. A unified or specified command with a broad continuing mission under a single commander established and so designated by the President, through the Secretary of Defense and with the advice and assistance of the Chairman of the Joint Chiefs of Staff. Also called **CCMD.** (Approved for incorporation into JP 1-02.)

combatant command (command authority). Nontransferable command authority, which cannot be delegated, of a combatant commander to perform those functions of command over assigned forces involving organizing and employing commands and forces; assigning tasks; designating objectives; and giving authoritative direction over all aspects of military operations, joint training, and logistics necessary to accomplish the missions assigned to the command. Also called **COCOM.** (Approved for incorporation into JP 1-02.)

command. 1. The authority that a commander in the armed forces lawfully exercises over subordinates by virtue of rank or assignment. 2. An order given by a commander; that is, the will of the commander expressed for the purpose of bringing about a particular action. 3. A unit or units, an organization, or an area under the command of one individual. Also called **CMD.** (Approved for incorporation into JP 1-02.)

command and control. The exercise of authority and direction by a properly designated commander over assigned and attached forces in the accomplishment of the mission. Also called **C2.** (Approved for incorporation into JP 1-02.)

GL-5

Glossary

command relationships. The interrelated responsibilities between commanders, as well as the operational authority exercised by commanders in the chain of command; defined further as combatant command (command authority), operational control, tactical control, or support. (JP 1-02. SOURCE: JP 1)

component. 1. One of the subordinate organizations that constitute a joint force. (JP 1) 2. In logistics, a part or combination of parts having a specific function, which can be installed or replaced only as an entity. (JP 4-0) Also called **COMP.** (Approved for incorporation into JP 1-02.)

continental United States. United States territory, including the adjacent territorial waters, located within North America between Canada and Mexico. Also called **CONUS.** (Approved for incorporation into JP 1-02 with JP 1 as the source JP.)

contingency operation. A military operation that is either designated by the Secretary of Defense as a contingency operation or becomes a contingency operation as a matter of law (Title 10, United States Code, Section 101[a][13]). (Approved for incorporation into JP 1-02.)

Contingency Planning Guidance. Secretary of Defense written guidance, approved by the President, for the Chairman of the Joint Chiefs of Staff, which focuses the guidance given in the national security strategy and Defense Planning Guidance, and is the principal source document for the Joint Strategic Capabilities Plan. Also called **CPG.** (Approved for incorporation into JP 1-02.)

continuity of command. None. (Approved for removal from JP 1-02.)

control. 1. Authority that may be less than full command exercised by a commander over part of the activities of subordinate or other organizations. (JP 1-02. SOURCE: JP 1) 2. In mapping, charting, and photogrammetry, a collective term for a system of marks or objects on the Earth or on a map or a photograph, whose positions or elevations (or both) have been or will be determined. (JP 1-02. SOURCE: JP 2-03) 3. Physical or psychological pressures exerted with the intent to assure that an agent or group will respond as directed. (JP 1-02. SOURCE: JP 3-0) 4. An indicator governing the distribution and use of documents, information, or material. Such indicators are the subject of intelligence community agreement and are specifically defined in appropriate regulations. (JP 1-02. SOURCE: JP 2-01)

coordinating authority. The commander or individual who has the authority to require consultation between the specific functions or activities involving forces of two or more Services, joint force components, or forces of the same Service or agencies, but does not have the authority to compel agreement. (Approved for incorporation into JP 1-02.)

defense readiness condition. None. (Approved for removal from JP 1-02.)

GL-6   JP 1

Glossary

delegation of authority. The action by which a commander assigns part of his or her authority, commensurate with the assigned task, to a subordinate commander. (Approved for incorporation into JP 1-02.)

Department of Defense components. The Office of the Secretary of Defense, the Military Departments, the Chairman of the Joint Chiefs of Staff and the Joint Staff, the combatant commands, the Office of the Inspector General of the Department of Defense, the Department of Defense agencies, Department of Defense field activities, and all other organizational entities in the Department of Defense. (Approved for incorporation into JP 1-02.)

Department of the Air Force. The executive part of the Department of the Air Force at the seat of government and all field headquarters, forces, Reserve Component, installations, activities, and functions under the control or supervision of the Secretary of the Air Force. Also called **DAF.** (Approved for incorporation into JP 1-02.)

Department of the Army. The executive part of the Department of the Army at the seat of government and all field headquarters, forces, Reserve Component, installations, activities, and functions under the control or supervision of the Secretary of the Army. Also called **DA.** (Approved for incorporation into JP 1-02.)

Department of the Navy. The executive part of the Department of the Navy at the seat of government; the headquarters, United States Marine Corps; the entire operating forces of the United States Navy and of the United States Marine Corps, including the Reserve Component of such forces; all field activities, headquarters, forces, bases, installations, activities, and functions under the control or supervision of the Secretary of the Navy; and the United States Coast Guard when operating as a part of the Navy pursuant to law. Also called **DON.** (Approved for incorporation into JP 1-02.)

directive authority for logistics. Combatant commander authority to issue directives to subordinate commanders to ensure the effective execution of approved operation plans, optimize the use or reallocation of available resources, and prevent or eliminate redundant facilities and/or overlapping functions among the Service component commands. Also called **DAFL.** (Approved for incorporation into JP 1-02.)

direct liaison authorized. That authority granted by a commander (any level) to a subordinate to directly consult or coordinate an action with a command or agency within or outside of the granting command. Also called **DIRLAUTH.** (Approved for incorporation into JP 1-02.)

economic action. None. (Approved for removal from JP 1-02.)

executive agent. A term used to indicate a delegation of authority by the Secretary of Defense or Deputy Secretary of Defense to a subordinate to act on behalf of the Secretary of Defense. Also called **EA.** (Approved for incorporation into JP 1-02.)

GL-7

Glossary

force. 1. An aggregation of military personnel, weapon systems, equipment, and necessary support, or combination thereof. 2. A major subdivision of a fleet. (JP 1-02. SOURCE: JP 1)

function. The broad, general, and enduring role for which an organization is designed, equipped, and trained. (Approved for replacement of "functions" and its definition in JP 1-02.)

functional component command. A command normally, but not necessarily, composed of forces of two or more Military Departments which may be established across the range of military operations to perform particular operational missions that may be of short duration or may extend over a period of time. (JP 1-02. SOURCE: JP 1)

general orders. None. (Approved for removal from JP 1-02.)

grand strategy. None. (Approved for removal from JP 1-02.)

inactive duty training. Authorized training performed by a member of a Reserve Component not on active duty or active duty for training and consisting of regularly scheduled unit training assemblies, additional training assemblies, periods of appropriate duty or equivalent training, and any special additional duties authorized for Reserve Component personnel by the Secretary concerned, and performed by them in connection with the prescribed activities of the organization in which they are assigned with or without pay. Also called **IDT.** (Approved for incorporation into JP 1-02.)

instruments of national power. All of the means available to the government in its pursuit of national objectives. They are expressed as diplomatic, economic, informational, and military. (JP 1-02. SOURCE: JP 1)

in support of. Assisting or protecting another formation, unit, or organization while remaining under original control. (JP 1-02. SOURCE: JP 1)

integration. 1. In force protection, the synchronized transfer of units into an operational commander's force prior to mission execution. 2. The arrangement of military forces and their actions to create a force that operates by engaging as a whole. 3. In photography, a process by which the average radar picture seen on several scans of the time base may be obtained on a print, or the process by which several photographic images are combined into a single image. (JP 1-02. SOURCE: JP 1)

irregular warfare. A violent struggle among state and non-state actors for legitimacy and influence over the relevant population(s). Also called **IW.** (Approved for incorporation into JP 1-02.)

joint. Connotes activities, operations, organizations, etc., in which elements of two or more Military Departments participate. (JP 1-02. SOURCE: JP 1)

Glossary

joint force commander. A general term applied to a combatant commander, subunified commander, or joint task force commander authorized to exercise combatant command (command authority) or operational control over a joint force. Also called **JFC.**
(JP 1-02. SOURCE: JP 1)

joint staff. 1. The staff of a commander of a unified or specified command, subordinate unified command, joint task force, or subordinate functional component (when a functional component command will employ forces from more than one Military Department), that includes members from the several Services comprising the force. 2. (capitalized as Joint Staff) The staff under the Chairman of the Joint Chiefs of Staff that assists the Chairman and the other members of the Joint Chiefs of Staff in carrying out their responsibilities. Also called **JS.** (Approved for incorporation into JP 1-02.)

joint task force. A joint force that is constituted and so designated by the Secretary of Defense, a combatant commander, a subunified commander, or an existing joint task force commander. Also called **JTF.** (JP 1-02. SOURCE: JP 1)

military characteristics. None. (Approved for removal from JP 1-02.)

Military Department. One of the departments within the Department of Defense created by the National Security Act of 1947, which are the Department of the Army, the Department of the Navy, and the Department of the Air Force. Also called **MILDEP.** (Approved for incorporation into JP 1-02.)

multinational force. A force composed of military elements of nations who have formed an alliance or coalition for some specific purpose. Also called **MNF.** (JP 1-02. SOURCE: JP 1)

national defense strategy. A document approved by the Secretary of Defense for applying the Armed Forces of the United States in coordination with Department of Defense agencies and other instruments of national power to achieve national security strategy objectives. Also called **NDS.** (Approved for incorporation into JP 1-02 with JP 1 as the source JP.)

national military strategy. A document approved by the Chairman of the Joint Chiefs of Staff for distributing and applying military power to attain national security strategy and national defense strategy objectives. Also called **NMS.** (Approved for replacement of "National Military Strategy" in JP 1-02.)

national policy. A broad course of action or statements of guidance adopted by the government at the national level in pursuit of national objectives. (Approved for incorporation into JP 1-02 with JP 1 as the source JP.)

national security. A collective term encompassing both national defense and foreign relations of the United States with the purpose of gaining: a. A military or defense advantage over any foreign nation or group of nations; b. A favorable foreign relations position; or c. A defense posture capable of successfully resisting hostile or destructive

GL-9

Glossary

action from within or without, overt or covert. (Approved for incorporation into JP 1-02.)

National Security Council. A governmental body specifically designed to assist the President in integrating all spheres of national security policy. Also called **NSC.** (Approved for incorporation into JP 1-02.)

national security interests. The foundation for the development of valid national objectives that define United States goals or purposes. (Approved for incorporation into JP 1-02.)

national security strategy. A document approved by the President of the United States for developing, applying, and coordinating the instruments of national power to achieve objectives that contribute to national security. Also called **NSS.** (Approved for replacement of "National Security Strategy" in JP 1-02.)

national support element. Any national organization or activity that supports national forces that are a part of a multinational force. (Approved for incorporation into JP 1-02.)

operation. 1. A sequence of tactical actions with a common purpose or unifying theme. (JP 1) 2. A military action or the carrying out of a strategic, operational, tactical, service, training, or administrative military mission. (JP 3-0) (Approved for incorporation into JP 1-02.)

operational authority. None. (Approved for removal from JP 1-02.)

operational control. The authority to perform those functions of command over subordinate forces involving organizing and employing commands and forces, assigning tasks, designating objectives, and giving authoritative direction necessary to accomplish the mission. Also called **OPCON.** (Approved for incorporation into JP 1-02.)

organic. Assigned to and forming an essential part of a military organization as listed in its table of organization for the Army, Air Force, and Marine Corps, and are assigned to the operating forces for the Navy. (Approved for incorporation into JP 1-02.)

other government agency. None. (Approved for removal from JP 1-02.)

partner nation. A nation that the United States works with in a specific situation or operation. Also called **PN.** (Approved for incorporation into JP 1-02.)

proper authority. None. (Approved for removal from JP 1-02.)

readiness. The ability of military forces to fight and meet the demands of assigned missions. (Approved for incorporation into JP 1-02.)

readiness condition. None. (Approved for removal from JP 1-02.)

Glossary

resources. The forces, materiel, and other assets or capabilities apportioned or allocated to the commander of a unified or specified command. (Approved for incorporation into JP 1-02 with JP 1 as the source JP.)

responsibility. None. (Approved for removal from JP 1-02.)

Secretary of a Military Department. None. (Approved for removal from JP 1-02.)

Service. A branch of the Armed Forces of the United States, established by act of Congress, which are: the Army, Marine Corps, Navy, Air Force, and Coast Guard. (Approved for replacement of "Military Service" and its definition in JP 1-02.)

Service component command. A command consisting of the Service component commander and all those Service forces, such as individuals, units, detachments, organizations, and installations under that command, including the support forces that have been assigned to a combatant command or further assigned to a subordinate unified command or joint task force. (JP 1-02. SOURCE: JP 1)

special staff. None. (Approved for removal from JP 1-02.)

specified combatant command. A command, normally composed of forces from a single Military Department, that has a broad, continuing mission, normally functional, and is established and so designated by the President through the Secretary of Defense with the advice and assistance of the Chairman of the Joint Chiefs of Staff. (Approved for replacement of "specified command" and its definition in JP 1-02.)

strategic vulnerability. None. (Approved for removal from JP 1-02.)

subordinate command. A command consisting of the commander and all those individuals, units, detachments, organizations, or installations that have been placed under the command by the authority establishing the subordinate command. (JP 1-02. SOURCE: JP 1)

subordinate unified command. A command established by commanders of unified commands, when so authorized by the Secretary of Defense through the Chairman of the Joint Chiefs of Staff, to conduct operations on a continuing basis in accordance with the criteria set forth for unified commands. (Approved for incorporation into JP 1-02.)

subunified command. None. (Approved for removal from JP 1-02.)

support. 1. The action of a force that aids, protects, complements, or sustains another force in accordance with a directive requiring such action. 2. A unit that helps another unit in battle. 3. An element of a command that assists, protects, or supplies other forces in combat. (JP 1-02. SOURCE: JP 1)

tactical control. The authority over forces that is limited to the detailed direction and control of movements or maneuvers within the operational area necessary to

Glossary

accomplish missions or tasks assigned. Also called **TACON.** (Approved for incorporation into JP 1-02.)

tactical warning and attack assessment. None. (Approved for removal from JP 1-02.)

task. A clearly defined action or activity specifically assigned to an individual or organization that must be done as it is imposed by an appropriate authority. (Approved for inclusion in JP 1-02.)

territorial airspace. Airspace above land territory and internal, archipelagic, and territorial waters. (Approved for incorporation into JP 1-02.)

territorial waters. A belt of ocean space adjacent to and measured from the coastal states baseline to a maximum width of 12 nautical miles. (Approved for replacement of "territorial sea" and its definition in JP 1-02.)

theater. The geographical area for which a commander of a geographic combatant command has been assigned responsibility. (JP 1-02. SOURCE: JP 1)

training and readiness oversight. The authority that combatant commanders may exercise over assigned Reserve Component forces when not on active duty or when on active duty for training. Also called **TRO.** (Approved for incorporation into JP 1-02.)

transient. None. (Approved for removal from JP 1-02.)

transient forces. Forces that pass or stage through, or base temporarily within, the operational area of another command but are not under its operational control. (JP 1-02. SOURCE: JP 1)

unified action. The synchronization, coordination, and/or integration of the activities of governmental and nongovernmental entities with military operations to achieve unity of effort. (JP 1-02. SOURCE: JP 1)

unified combatant command. See unified command. (JP 1-02. SOURCE: JP 1)

unified command. A command with a broad continuing mission under a single commander and composed of significant assigned components of two or more Military Departments that is established and so designated by the President, through the Secretary of Defense with the advice and assistance of the Chairman of the Joint Chiefs of Staff. (JP 1-02. SOURCE: JP 1)

Unified Command Plan. The document, approved by the President, that sets forth basic guidance to all unified combatant commanders; establishes their missions, responsibilities, and force structure; delineates the general geographical area of responsibility for geographic combatant commanders; and specifies functional responsibilities for functional combatant commanders. Also called **UCP.** (JP 1-02. SOURCE: JP 1)

United States. Includes the land area, internal waters, territorial sea, and airspace of the United States, including a. United States territories; and b. Other areas over which the United States Government has complete jurisdiction and control or has exclusive authority or defense responsibility. (Approved for incorporation into JP 1-02.)

United States Armed Forces. Used to denote collectively the Army, Marine Corps, Navy, Air Force, and Coast Guard. (Approved for incorporation into JP 1-02.)

unity of effort. Coordination and cooperation toward common objectives, even if the participants are not necessarily part of the same command or organization, which is the product of successful unified action. (Approved for incorporation into JP 1-02.)

US forces. All Armed Forces (including the Coast Guard) of the United States, any person in the Armed Forces of the United States, and all equipment of any description that either belongs to the US Armed Forces or is being used (including Type I and II Military Sealift Command vessels), escorted, or conveyed by the US Armed Forces. (Approved for incorporation into JP 1-02 with JP 1 as the source JP.)

US national. US citizen and US permanent and temporary legal resident aliens. (Approved for incorporation into JP 1-02 with JP 1 as the source JP.)

Intentionally Blank

JOINT DOCTRINE PUBLICATIONS HIERARCHY

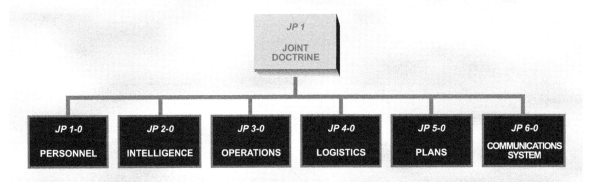

All joint publications are organized into a comprehensive hierarchy as shown in the chart above. **Joint Publication (JP) 1 is the capstone joint doctrine publication.** The diagram below illustrates an overview of the development process:

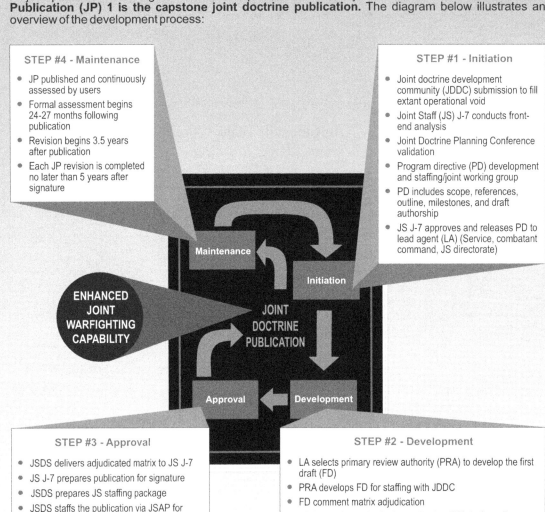

STEP #4 - Maintenance
- JP published and continuously assessed by users
- Formal assessment begins 24-27 months following publication
- Revision begins 3.5 years after publication
- Each JP revision is completed no later than 5 years after signature

STEP #1 - Initiation
- Joint doctrine development community (JDDC) submission to fill extant operational void
- Joint Staff (JS) J-7 conducts front-end analysis
- Joint Doctrine Planning Conference validation
- Program directive (PD) development and staffing/joint working group
- PD includes scope, references, outline, milestones, and draft authorship
- JS J-7 approves and releases PD to lead agent (LA) (Service, combatant command, JS directorate)

STEP #3 - Approval
- JSDS delivers adjudicated matrix to JS J-7
- JS J-7 prepares publication for signature
- JSDS prepares JS staffing package
- JSDS staffs the publication via JSAP for signature

STEP #2 - Development
- LA selects primary review authority (PRA) to develop the first draft (FD)
- PRA develops FD for staffing with JDDC
- FD comment matrix adjudication
- JS J-7 produces the final coordination (FC) draft, staffs to JDDC and JS via Joint Staff Action Processing (JSAP) system
- Joint Staff doctrine sponsor (JSDS) adjudicates FC comment matrix
- FC joint working group

Made in the USA
Las Vegas, NV
22 March 2021